IUTAM Symposium on Evolutionary Methods in Mechanics

SOLID MECHANICS AND ITS APPLICATIONS
Volume 117

Series Editor: G.M.L. GLADWELL
Department of Civil Engineering
University of Waterloo
Waterloo, Ontario, Canada N2L 3GI

Aims and Scope of the Series

The fundamental questions arising in mechanics are: *Why?, How?,* and *How much?*
The aim of this series is to provide lucid accounts written by authoritative researchers giving vision and insight in answering these questions on the subject of mechanics as it relates to solids.

The scope of the series covers the entire spectrum of solid mechanics. Thus it includes the foundation of mechanics; variational formulations; computational mechanics; statics, kinematics and dynamics of rigid and elastic bodies: vibrations of solids and structures; dynamical systems and chaos; the theories of elasticity, plasticity and viscoelasticity; composite materials; rods, beams, shells and membranes; structural control and stability; soils, rocks and geomechanics; fracture; tribology; experimental mechanics; biomechanics and machine design.

The median level of presentation is the first year graduate student. Some texts are monographs defining the current state of the field; others are accessible to final year undergraduates; but essentially the emphasis is on readability and clarity.

For a list of related mechanics titles, see final pages.

IUTAM Symposium on

Evolutionary Methods in Mechanics

Proceedings of the IUTAM Symposium held in
Cracow, Poland, 24–27 September, 2002.

Edited by

TADEUSZ BURCZYŃSKI

Department for Strength of Materials,
Silesian University of Technology, Gliwice, Poland
and
Institute of Computer Modelling,
Cracow University of Technology, Cracow, Poland

and

ANDRZEJ OSYCZKA

Department of Applied Computer Science,
Faculty of Management,
AGH University of Science and Technology,
Cracow, Poland

KLUWER ACADEMIC PUBLISHERS
DORDRECHT / BOSTON / LONDON

A C.I.P. Catalogue record for this book is available from the Library of Congress.

ISBN   978-90-481-6628-2 (PB)
ISBN   978-1-4020-2267-8 (e-book)

Published by Kluwer Academic Publishers,
P.O. Box 17, 3300 AA Dordrecht, The Netherlands.

Sold and distributed in North, Central and South America
by Kluwer Academic Publishers,
101 Philip Drive, Norwell, MA 02061, U.S.A.

In all other countries, sold and distributed
by Kluwer Academic Publishers,
P.O. Box 322, 3300 AH Dordrecht, The Netherlands.

Printed on acid-free paper

Contents

vi

Preface

The IUTAM Symposium on Evolutionary Methods in Mechanics was held in Cracow, Poland, September 24-27, 2002. The site of the Symposium was Cracow University of Technology. The Symposium was attended by 50 persons from 18 countries. In addition, several Polish students, Ph. D. students and research associates participated in the meeting.

The Symposium provided an excellent opportunity for scholars of mechanics, computer sciences and artificial intelligence to interact and exchange their points of view on the advanced computational and application aspects of the evolutionary methods in analysis and design of mechanical systems. Recently evolutionary methods have become the most effective tools for solving specific kinds of problems in mechanics, especially in structural and multidisciplinary optimization. The meeting was devoted to both theoretical and practical developments of computational mechanics methods drawing their inspiration from nature with particular emphasis on evolutionary models of computation such as genetic algorithms, evolutionary strategies, classifier systems, evolutionary programming and other evolutionary computation techniques in mechanics. The objective of the Symposium was to provide an international forum for facilitating the exchange of information among researchers involved in computational intelligence methods based on evolutionary nature. The Symposium put special emphasis on evolutionary optimization in various fields of mechanics. The subject of evolutionary optimization has recently experienced a remarkable growth. New concepts, approaches and applications are being continually developed and exploited to provide efficient tools for solving a variety of optimization problems in mechanics. The topics covered by the Symposium included:

- Evolutionary methods in shape and topology optimization,

- Evolutionary methods in size and material optimization,

- Evolutionary methods in multicriteria optimization,

- Evolutionary methods in inverse problems of engineering mechanics,

- Evolutionary methods in biomechanics,

- Coupling of evolutionary methods and neural networks in mechanics,

- Methods of artificial intelligence in mechanics based on evolutionary approaches,

- Other unconventional applications in mechanics.

The lectures and discussions about them clearly showed the remarkable progress in applications of evolutionary methods in mechanics. The volume contains 33 papers. Much to regret of the Scientific Committee some manuscripts were not submitted. All papers contained herein have been reviewed to the standard of leading scientific journals. The Editors would like to acknowledge the great efforts on behalf of both the authors and reviewers. The Editors particularly wish to thank the Bureau of the International Union of Theoretical and Applied Mechanics (IUTAM), the International Society of Structural and Multidisciplinary Optimization (ISSMO) and the Internal Scientific Committee. Part of success of the Symposium was a consequence of the excellent facilities provided by the Cracow University of Technology. The smooth running of the Symposium owes much to the initiative and the organizational skills of B.Ulejczyk, M.Dziewoński, J.Habel, A.Karafiat, S.Krenich, W.Kulig, W.Kuś, P.Orantek and A.Sobos. Finally, Editors would like to express their gratitude to the sponsoring organizations who have supported the Symposium financially, namely the IUTAM, the ISSMO, the Committee of Mechanics of Polish Academy of Sciences, the Polish Association for Computational Mechanics, Cracow University of Technology and Silesian University of Technology.

Cracow/Gliwice, February 2004
Tadeusz Burczyński
Andrzej Osyczka
Editors

Comittees and Sponsors

International Scientific Committee

T. Burczyński, Co-Chairman (Poland)

A. Osyczka, Co-Chairman (Poland)

K. Deb (India)

J.Engelbrecht (Estonia)

D.E.Grierson (Canada)

R.Haftka (USA)

P. Hajela (USA)

S. Kundu (Japan)

I.C. Parmee (UK)

Sponsors

International Union of Theoretical and Applied Mechanics (IUTAM)

International Society of Structural
and Multidisciplinary Optimization (ISSMO)

Polish Association for Computational Mechanics (PACM)

Committee of Mechanics of Polish Academy of Sciences

Department for Strength of Materials and Computational Mechanics,
Silesian University of Technology, Gliwice, Poland

Institute of Computer Modelling, Artificial Intelligence Division
Institute of Production Engineering,
Cracow University of Technology, Cracow, Poland

EVOLUTIONARY COMPUTATION IN CRACK PROBLEMS

Witold Beluch
Department for Strength of Materials and Computational Mechanics, Silesian University of Technology, Konarskiego 18a, PL-44-100 Gliwice, Poland

Abstract: Evolutionary algorithms (EA) seem to be very interesting optimization algorithms, sometimes more efficient than the classical (especially gradient) ones. The application of EAs for many different problems connected with the optimization in fracture mechanics: the shape optimization of cracked structures, the identification of the cracks, the identification of the boundary conditions in cracked structures, the prediction of the crack growth path is presented. Boundary element method is used for solving the boundary-value problem. Numerical tests are included.

Keywords: evolutionary algorithm, fracture mechanics, optimization, identification, boundary element method

1.   INTRODUCTION

In the present paper the application of evolutionary algorithms (EAs) for solving problems connected with cracked mechanical structures is presented, namely:
- the shape optimization of the external boundary;
- the identification of the shape, position and number of the cracks;
- the identification of the boundary conditions.

All the foregoing problems could be formulated as the optimization ones. There are many different optimization methods, especially based on the gradient of the objective function [7]. Although they are very accurate and fast, there are many inconveniences: objective function must be continuous, the probability of convergence to a local optimum is very large, etc. EAs are

T. Burczyński and A. Osyczka (eds),
IUTAM Symposium on Evolutionary Methods in Mechanics, 1–11.
© *2004 Kluwer Academic Publishers.*

optimization methods which enable avoiding many inconveniences connected with gradient methods, especially when the objective function is multi-
modal and the value of its gradient is difficult to obtain. In order to calculate
the objective function, it is essential to solve a boundary-value problem. In
the present paper the boundary element method has been employed.

2.   PROBLEM FORMULATION

2.1   Optimization problem

The optimization problem is formulated as the minimization of the objective function J_0:

$$\min : J_0(x) \tag{1}$$

with constrains:

$$
\begin{aligned}
&J_\alpha(x){=}0, \quad \alpha{=}1,2,...,n \\
&J_\beta(x) \,\square\, 0, \quad \beta{=}1,2,...,n \\
&x_{i_{max}} \,\square\, x_i \,\square\, x_{i_{min}}, \quad i{=}1,2,...,k
\end{aligned} \tag{2}
$$

where: $J_\square$, $J_\square$ – constrain functionals; x – shape design variables vector;
n, m, k – constants.
Depending on the problem, the objective functions are defined in the forms
presented in Chapters 3-5.
In order to calculate the objective functional value, one has to solve a boundary-value problem. This problem can be solved by means of the finite
element method (FEM) or the boundary element method (BEM). Since the
crack is a part of the boundary, the BEM seems to be the most convenient
one.

2.2   Boundary Element Method in fracture mechanics

An elastic body occupying a domain $\square$ and having a boundary $\square\square\square$ is
considered (Fig.1). Two fields are prescribed on the boundary $\square$: a field of
displacements $u^o(x)$, $x\square\square_u$, and a field of tractions $p^o(x)$, $x\square\square_p$, and $\square_u\square\square_p =$
$\square$ and $\square_u\square\square_p = \square$. The body contains internal traction-free cracks C_i.

Displacements are allowed to jump across C:

$$[\![u]\!] \Box u^{+} - u^{-} \Box 0 \tag{3}$$

The BEM seems to be the most suitable method for the problems with cracks because it is capable of accurate modelling the high stress gradients near the crack tip and (assuming the lack of body forces) the discretization of the inside of the body is not required.

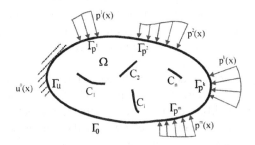

Figure 1. An elastic body containing cracks

When the body forces are neglectable, the displacement of a point **x** can be represented by the following boundary displacement integral equation:

$$\mathbf{c(x)u(x)} = \int \mathbf{U(x,y)p(y)}d\Box\mathbf{(y)} \Box \int \mathbf{P(x,y)u(y)}d\Box\mathbf{(y)}, \quad \mathbf{x}\Box\Box \tag{4}$$

where: **U(x,y)**, **P(x,y)** – fundamental solutions of elastostatics; **c(x)** – a constant depending on the collocation point (**x**) position; **y** – the boundary point.

Applying eq. (4) on both crack surfaces two identical equations are created, so the consequential set of algebraic equations is singular. In order to overcome this difficulty the technique called *the dual boundary integral equation method* [6] is used. In this technique two different boundary equations are applied on both sides of the crack, namely equation (4) and a hypersingular tractions integral equation in the form:

$$\tfrac{1}{2}\mathbf{p(x)} = \mathbf{n}\left[\int \mathbf{D(x,y)p(y)}d\Box\mathbf{(y)} \Box \int \mathbf{S(x,y)u(y)}d\Box\mathbf{(y)} \right], \quad \mathbf{x}\Box\Box \tag{5}$$

where: **D(x,y)**, **S(x,y)** - the third-order fundamental solution tensors, **n** - the unit outward normal vector at the collocation point **x**.

The tractions integral equation is applied on one surface of each crack; the displacements integral equation is applied on the opposite side of each crack and the remaining boundary.

2.3   Evolutionary program

Evolutionary algorithms originate from classical genetic algorithms (GAs), which are stochastic algorithms modelling natural phenomena: genetic inheritance and Darwinian strife for survival. GAs work on populations of solutions and are based only on the fitness (objective) function value information. In the classical GAs binary coding is used, binary operators of crossover and mutation are employed and the probability of operators is constant [4].

Evolutionary algorithms (EAs) are modified and generalized classical genetic algorithms in which different representations of the population (usually a floating point) and modified operators are used. The selection is usually performed in the form of the ranking selection or the tournament selection. The probability of operators can be variable.

The evolutionary program used in the present work consists of two major blocks: the evolutionary algorithm block and the objective function evaluation block (Fig. 2).

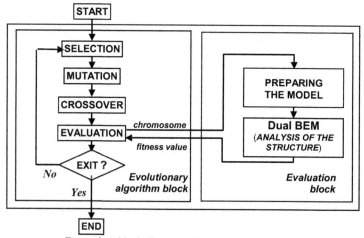

Figure 2. A block diagram of the evolutionary program

The floating point gene representation and the ranking selection are used. Five evolutionary operators are applied: uniform and boundary mutations, simple, arithmetical and heuristic crossovers [5].

3.   SHAPE OPTIMIZATION

In the optimization of the external shape of cracked structures several criteria are proposed [1]:
a) the minimization of the maximum crack opening (*MCO*):

$$J_0 = MCO_{red} = \sum_{i=1}^{n} w_i MCO_i \qquad (6)$$

where: $MCO = \max(u^+ - u^-)$; u^+, u^- – the displacement values of the coincident nodes laying on the opposite sides of the crack; $w_i = MCO_i/\Box MCO_i$ – weight factors ($w_i = 1$); n – number of cracks.
b) the minimization of the reduced J-integral:

$$J_0 = J = \sum_{i=1}^{n} w_i J_i \qquad (7)$$

where: J_i – J-integral for i-tip of the crack; $w_i = J_i/\Box J_i$; $n = 2 \cdot$ number of cracks.
c) the minimization of the reduced stress intensity factor in the form:

$$J_0 = K_{red} = \sum_{i=1}^{n} w_i K_i \qquad (8)$$

where: K_i – stress intensity factors; $w_i = K_i/\Box K_i$; $n = 4 \cdot$ number of cracks.
Traction-free and unconstraint parts of the external boundary are modified. The restriction for the maximum value of the boundary von Misses reduced stresses is employed.

3.1   Numerical examples

3.1.1   Shape optimization - example 1

The boundary of a 2D structure containing two cracks (Fig. 3a) is optimized. The objective of the optimization is to minimize K_{red}. Constraints on the equivalent von Misses stresses are imposed on the boundary. The optimal shape is presented in Fig. 3b. K_{red} has been reduced from 4.2679 to 3.9072.

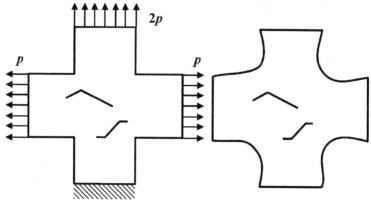

Figure 3. A structure with 2 cracks a) before, b) after optimization

3.1.2   Shape optimization - example 2

A structure containing three linear cracks (Fig. 4a) is optimized and the criterion of minimum MCO_{red} is applied. Constraints on the equivalent von Misses stresses are imposed on the boundary. The optimal shape is presented in Fig. 4b. MCO_{red} has been reduced from 1.8389 to 1.53386.

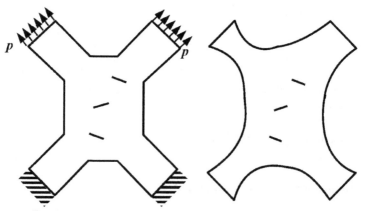

Figure 4. A structure with 3 cracks a) before, b) after optimization

4.   CRACKS IDENTIFICATION

A crack identification problem [3] can be presented as the minimization of the objective function J_0 with respect to $\mathbf{a}=(a_i)$:

$$\min_{\mathbf{a}}(J_0) = \min_{\mathbf{a}}\left\{ \int_{\Gamma_0} \mathbf{W}(\mathbf{x})[\mathbf{q}(\mathbf{x}) \square \hat{\mathbf{q}}(\mathbf{x})]^2 \, d\square(\mathbf{x}) \right\} \tag{6}$$

where: $\hat{\mathbf{q}}_j$ – "measured" values of a boundary state field (displacements or stresses); $\mathbf{q}_j$ – computed values of the same state field; $\mathbf{a}$ – parameters representing the number, the position and the shape of the cracks .

The following problems are considered:
– the identification of the single cracks;
– the identification of the multiple cracks with known cracks number;
– the identification of the multiple cracks with unknown cracks number;
– the identification of the cracks with the stochastic disturbance of the
  measurements.

One assumes that measured state fields have stochastic nature and are characterized by the mean value $E(\hat{\mathbf{q}})$ and the standard deviation $D(\hat{\mathbf{q}})$.

4.1   Numerical examples

4.1.1   Cracks identification - example 1

The problem of the identification of a crack when the displacements measured at 41 boundary (sensor) points were disturbed by random fluctuations was considered. It is assumed that $D(\hat{\mathbf{q}}) = E(\hat{\mathbf{q}})/30$. The final and actual position of the crack is presented in Fig. 5.

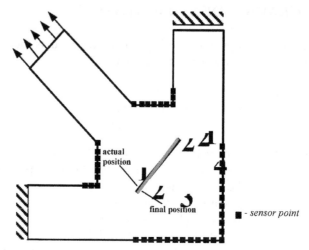

Figure 5. The identification of a crack with measurement error

4.1.2   Cracks identification - example 2

It is assumed that there are $1|5$ linear cracks of unknown lengths and positions inside the structure. The actual number of the cracks is 2. The aim is to find the number and positions of the cracks. Displacements at 81 sensor points are measured. The actual number of cracks was successfully found. The actual and final positions of the cracks are presented in Fig. 6.

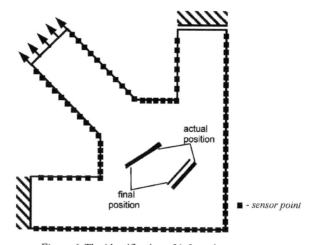

Figure 6. The identification of 1-5 cracks

5. IDENTIFICATION OF THE BOUNDARY CONDITIONS

The identification of the boundary conditions in the form of tractions or concentrated loads can be expressed by (6) where **a** are the parameters representing the distribution of the tractions or concentrated loads. The subsequent problems are considered [2]:
- the identification of the boundary condition values;
- the identification of the boundary condition values and positions;
- the identification of the boundary condition values, positions and numbers.

The influence of the stochastic disturbance of the measurements has been considered as well.

5.1 Numerical examples

5.1.1 Identification of the boundary conditions - example 1

A 2-D structure containing a crack (Fig. 4) is considered. The aim is to identify the values of 3 tractions p_1, p_2 and p_3 having measured displacements at sensor points. The measurement error ("noise") is considered as well. Results are collected in Table 1.

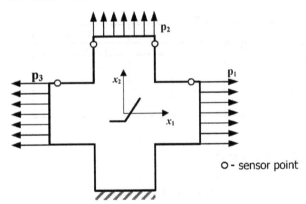

Figure 7. The identification of 3 tractions

Table 1. The identification of 3 tractions

Traction			No noise		Noise	
No.	Limitations	Actual value	Final value	Error [%]	Final value	Error [%]
1	-50; 50	15.0	15.7546	5.03	15.4583	3.06
2	-100; 100	10.0	10.0484	0.48	10.4867	4.87
3	-50; 50	15.0	15.7516	5.01	15.4428	2.95

5.1.2   Identification of the boundary conditions - example 2

A 2-D structure containing a crack (Fig. 5) is considered. The aim is to identify the number and the values of tractions having measured displacements at sensor points. There are 5 possible non-zero tractions (p1-p5). It is assumed, that if the generated traction value is less than 1.0, this traction does not exist. In fact there are 3 non-zero tractions. The measurement error is also taken into account. Results are presented in Table 2.

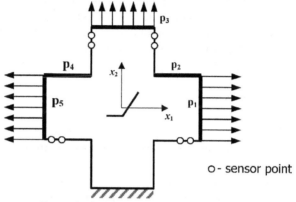

Figure 8. The identification of 0-5 tractions

Table 2. The identification of 0-5 tractions

Traction			No noise		Noise	
No.	Limitations	Actual value	Final value	Error [%]	Final value	Error [%]
1	0; 20	15.0				
2	0; 20	0.0				
3	0; 20	10.0				
4	0; 20	0.0				
5	0; 20	15.0				

6.   CONCLUSIONS

In the present paper evolutionary algorithm has been joined with the dual boundary element method into the evolutionary program. This attitude appears to be very useful in the optimization and identification problems of the cracked mechanical structures. The inverse problems, like identification, are very good for testing the used algorithm because the exact solution is known.

Evolutionary program is especially efficient when the number of cracks or tractions (forces) is unknown. It is also very error-resistant - the stochastic disturbance of measurements does not change the results critically, though the computation can be more time-consuming.

Using the parametric curves in the boundary optimization tasks allows decreasing the number of design variables.

ACKNOWLEDGEMENT

The support from KBN grant no. 4T11F00822 is gratefully acknowledged.

References

[1] W. Beluch. *Sensitivity analysis and evolutionary optimization of cracked mechanical structures.* Ph.D. thesis, Mechanical Engineering Faculty, Silesian University of Technology, Gliwice, 2000 (in Polish).

[2] W. Beluch, T. Burczyński *Identification and optimization of boundary conditions using evolutionary algorithms.* Proc. AI-MECH 2001 Symposium on Methods of Artificial Intelligence in Mechanics and Mechanical Engineering, Gliwice, Poland 2001.

[3] T. Burczyński, W. Beluch, A. Długosz, P. Orantek and M. Nowakowski. *Evolutionary methods in inverse problems of engineering mechanics.* Proc. International Symposium on Inverse Problems in Engineering Mechanics - ISIP 2000, Nagano (Japan), Elsevier London, 2000.

[4] D.E. Goldberg. *Algorytmy genetyczne i ich zastosowania*, WNT, Warszawa, 1995

[5] Z. Michalewicz. *Genetic Algorithms + Data Structures = Evolutionary Programs.* Springer-Verlag, AI Series, New York, 1992.

[6] A. Portela, M.H. Aliabadi and D.P. Rooke. *The dual boundary element method: effective implementation for crack problems.* Int. J. Num. Meth. Eng., No.33,1992, pp.1269-1287.

[7] D.R. Smith. *Variational Methods In Optimization*, Dover Publications, Inc., Mineola, New York, 1998.

INVESTIGATION OF EVOLUTIONARY ALGORITHM EFFECTIVENESS IN OPTIMAL SYNTHESIS OF CERTAIN MECHANISM

Krzysztof Białas-Heltowski, Tomasz Klekiel, Ryszard Rohatyński
Department of Mechanical Engineering, University of Zielona Góra, Podgórna 50, 62-246 Zielona Góra, Poland

Abstract: This paper presents the comparison of four evolutionary algorithms (EA) in terms of their effectiveness as applied to solve an engineering optimization task. The four tested algorithms differed in structure. The problem to be solved concerned optimization of a four-bar linkage dimensions so that its movement follows a given trajectory. Best EA parameters values were sought, namely: crossover probability, mutation probability and pool size. The main EA evaluation criterion was the number of iterations necessary to set the size of a mechanism that would realize the trajectory with a given tolerance. Substantial dispersion of results obtained from the algorithms after multiple repetitions of calculations were observed. For that reason the second criterion was introduced, a statistical measure of iterations number dispersion. The combined criteria, the number of iterations and standard deviation, enabled different EAs effectiveness comparison and indicated proper selection of their parameters values.

Keywords: evolutionary algorithm, optimization, mechanism synthesis

1.   EVOLUTIONARY ALGORITHMS TESTED

In optimization problems that involve complicated and time-consuming calculations of fitness function, the number of iterations necessary to achieve the satisfying solution should be possibly small. Consequently, in such problems the selection of an adequate evolutionary algorithm and its parameters is of high importance. However, only the number of iterations

13

T. Burczyński and A. Osyczka (eds),
IUTAM Symposium on Evolutionary Methods in Mechanics, 13–22.
© *2004 Kluwer Academic Publishers.*

may not be sufficient a criterion. Evolutionary algorithms include probability mechanisms, which - in general - after the repetition of calculations do not guarantee achieving the required result with the same number of iterations. The number of iterations needed to find a satisfactory solution can be a sufficient criterion for the algorithm only if its repetitions show minor differences. But with high dispersion of this number the probability to find the optimal solution in one attempt highly decreases.

The authors have proved that to recommend the number of iterations, it is necessary to allow for the dispersion of results. The dispersion can be evaluated through multiple repetition of the calculation with the unchanged input data and the same values of parameters adjusting the algorithm performance. In this paper the standard deviation has been applied as a statistical measure for the number of iterations necessary to yield a satisfying solution. Investigation of the influence of AE parameters on this indicator showed that it is statistically stable, that is the iterations number dispersion is a feature of AE.

Four evolutionary algorithms were adapted to experiments, two of which have been taken from the textbooks with no changes in the evolutionary algorithm core part; they were only adopted for the experiment's needs. The source code of those programmes has been modified so that it was possible automatically to carry out the calculations for various sets of parameters, and to interface with programmes calculating the fitness function. The third algorithm differs from the second one in its mode for crossover, the fourth one has been developed by the second author. To sum up:

a)   AE I is virtually a genetic algorithm SGA, taken from [2] and extended for multidimensional optimization.

b)   AE II is an evolutionary algorithm from [3].

c)   AE III is modified AE II by replacing a changing crossover by averaging crossover [1].

d)   AE IV has been originally devised by the second author in [6].

2.   OPTIMIZATION PROBLEM CASE

The geometry of a four-bar linkage was sought for that five positions $E_1F_1 \div E_5F_5$ of its EF member was predetermined (Fig. 1.). Distances EB and FC as well as angles α and β should be determined so that EF member followed the given positions as exactly as possible.

For the optimization task two objective functions, f1, and f2 were accepted. If we assume a tentative mechanism follows exactly predetermined five EF positions, then the objective functions f1 and f2 are the optimal ones.

If a not optimal mechanism still follows the given trajectory, then the base pivotal points A and D must move and take three positions, as shown in Fig.2. Consequently, the objective function f1 is a measure of the distance between the triangle vertexes and its center. The base points A, B will be fixed only for the optimal mechanism.

The second, (f2), function is simply a measure of the deviation of the actual EF points from the predetermined ones. The functions f1 and f2 play the role of the fitness functions of the evolutionary algorithms. After having calculated f1 and f2, if a roulette wheel selection system is used, one can evaluate fitness values of the population members. If the tournament selection is applied, the comparison of the f1 and f2 values indicates, which chromosome passes to the next population.

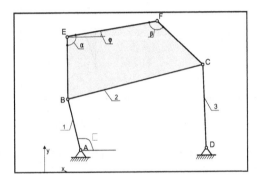

Figure 1. The optimized mechanizm

For given positions $E_1F_1 \div E_5F_5$ and any tentative values of the four optimization variables α, β, EB, FC, the base points, $A(x_a, y_a)$ and $D(x_d, y_d)$ coordinates and the members AB, BC, and CD length can be calculated as follows:

a) Successive positions of B_i and C_i points, where i □ <1;5>, for a given α, β, EB, FC; $E_1F_1 \div E_5F_5$ are determined by the following relations:

$$B_i(x_{Bi}; y_{Bi}) = \begin{cases} x_{Bi} = x_{Ei} + |EB| \cos(\square + \square) \\ y_{Bi} = y_{Ei} + |EB| \sin(\square + \square) \end{cases} \quad C_i(x_{Ci}; y_{Ci}) = \begin{cases} x_{Ci} = x_{Fi} + |FC| \cos(\square + \square) \\ y_{Ci} = y_{Fi} + |FC| \sin(\square + \square) \end{cases} \quad (1)$$

Of course, BC length is constant for any fixed values of optimization variables.

b) Having determined $B_1B_5 \div C_1C_5$ coordinates we draw four bisectors from straight segments $B_1B_5 \div C_1C_5$. The intersections of the neighbouring bisectors define coordinates of the A_j and D_j points; i □<1;3> (Fig. 2.)

c) The averaged coordinates of pivots A and D are calculated by the relations:

$$A(x_A, y_A) = \frac{1}{3}\sum_{j=1}^{3} A_j(x_{Aj}, y_{Aj}) \qquad D(x_D, y_D) = \frac{1}{3}\sum_{j=1}^{3} D_j(x_{Dj}, y_{Dj}) \tag{2}$$

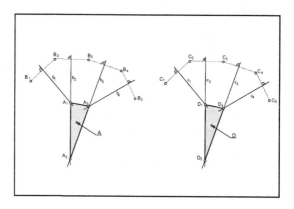

Figure 2. Construction of the averaged base pivotal points

d) Sum of distances between points A and Ai and D and Di is the fitness measure (f1):

$$f1 = \sum_{i=1}^{3} \left(\left| \overline{AA_i} \right| + \left| \overline{DD_i} \right| \right) \tag{3}$$

e) For further calculations the averaged length of links AB and DC are taken:

$$\left| \overline{AB} \right| = \frac{1}{5}\sum_{j=1}^{5} \sqrt{(x_A \Box x_{Bj})^2 + (y_A \Box y_{Bj})^2} \qquad \left| \overline{DC} \right| = \frac{1}{5}\sum_{j=1}^{5} \sqrt{(x_D \Box x_{Cj})^2 + (y_D \Box y_{Cj})^2} \tag{4}$$

f) Now trajectories T_E and T_F of the E and F points can be calculated. It is assumed that AB link is the crank and DC is the rocker. The crank can rotate in the range $\delta \Box <0;2\pi>$, where δ is AB inclination angle to X-axis [4, 5].

g) The second fitness function f2 is evaluated by means of distances between the given and calculated EF points. The problem of finding the points in calculated trajectories T_E and T_F that correspond to the assumed EF points can be resolved by means of elementary geometry.

3.   RESULTS OF EXPERIMENTS

In this section we discuss the results of the application of four previously mentioned algorithms in the task explained in section 2. It had been assumed that the main algorithm effectiveness indicator would be the number of iterations necessary to achieve the required level of the mechanism quality measure. The second criterion is the dispersion of the iterations acquired in 10 calculation tests. One should notice, that the number of iterations does not exactly match with the time of calculations, as the second also depends on the population size (PS). The PS value then, may also be taken into account fin the algorithm evaluation.

To economize the number of experiments a Latin Square plan has been devised. Assumed values of EA parameters are shown in Table 1. The abbreviation PS stands for number of population; PC marks the crossover probability, and PM mutation probability.

Table 1. Values of the EA parameters for numerical experiments

Lp	PS	PC	PM
1	30	0,5	0,01
2	50	0,6	0,05
3	80	0,7	0,1
4	100	0,8	0,15
5	120	0,9	0,2

For this number of variables and this number of their values it is possible to construct 4 experiment plans. The Latin square chosen is shown in Table 2. The figures in the first column indicate PS values. The figures in the first row – the PC values (from Table 1). The numbers at the crossing of a row and a column identifies the PM values (from Table 1). To sum up, there are 25 sets of parameters, which have been applied for four evolutionary algorithms tests.

Table 2. The Latin Square for numerical experiment

	1	2	3	4	5
1	1	2	3	4	5
2	3	4	5	1	2
3	5	1	2	3	4
4	2	3	4	5	1
5	4	5	1	2	3

For all algorithms the ultimate value 0,1 of the fitness function f2 was assumed as the stopping citerion. This value was calculated from the shift between the given and the actually performed trajectories. If the program had not reached this value in 2000 iterations it also stopped.

It has been observed that although all of the tested algorithms achieved the fitness function required value, the numbers of iterations though depended not only on the type of algorithm and its set up, but they were also different in repetitions. That led to the check of their dispersion. In order to do that the calculations for each set of parameters were repeated 10 times. The results are shown on Figures 3 to 6.

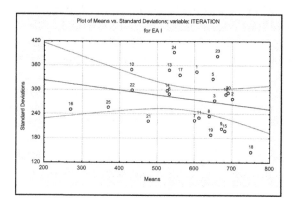

Figure 3. Mean number and dispersion of iterations for EA I with particular sets of parameters

Figure 3 outlines the results of the experiment for the EA I algorithm. The most promising settings are 16 and 18, with the corresponding parameters EA: PS=100, PC=0.5, PM=0.05, and PS=100, PC=0.7, PM=0.15, respectively. These sets differ as well in terms of mean iterations number as the dispersion. According to individual user preferences either 16 or 18 may be declared better.

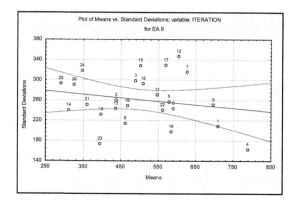

Figure 4. Mean number and dispersion of iterations for EA II with particular sets of parameters

Figure 4 presents the results for EA II algorithm experiment. Taking into account iterations number dispersion and its mean value two settings, number 4 and 23, can be distinguished, with the following corresponding parameters EA: PS=30, PC=0.8, PM=0.15, and PS=120, PC=0.7, PM=0.01. Iterations number dispersion is smaller in set 4 than in set 23, but the latter requires fewer number of iterations.

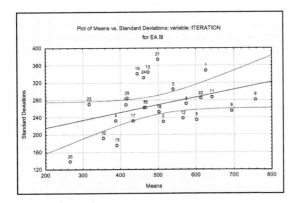

Figure 5. Mean number and dispersion of iterations for EA III with particular sets of parameters

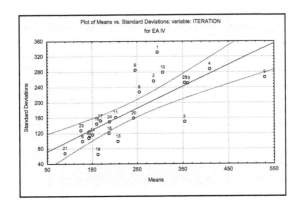

Figure 6. Mean number and dispersion of iterations for EA IV with particular sets of parameters

Figure 5 shows the results for EA III algorithm. In this case the best set is set 20 with EA parameters: PS=100, PC=0.9, PM=0.01. It features the smallest iterations dispersion and at the same time the lowest mean value.

Figure 6 shows the results for algorithm EA IV. Here the best set is 21 with the following values of EA parameters: PS=120, PC=0.5, PM=0.15. Set 6 also marks out, with its EA parameters: PS=50, PC=0.5, PM=0.1. The differences between these two sets are minor, however, because of the fact that in set 6 the pool size is two times smaller will probably shorten the calculation time, which may be advantageous despite the higher number of iterations.

Next, the probability of reaching the required fitness function limit for the best sets of EA parameters was defined. Having approved that the acquired results could be approximated with normal distribution, the numbers of iterations were assigned of the adequate distribution function. The results are shown in Figures 7 to10.

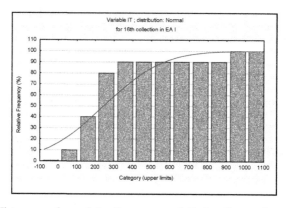

Figure 7. Histogram and cumulative frequency graph for iteration numbers at set 16 (AE I)

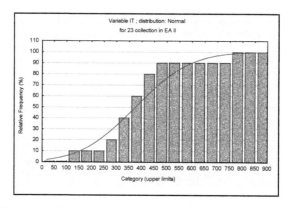

Figure 8. Histogram and cumulative frequency graph for iteration numbers at set 23 (EA II)

According to these figures, algorithm EA I features the lowest efficiency, as it requires the highest number of iterations, thus the solution seeking time is the longest. Replacing binary coding with the floating point coding slightly enhances the algorithm efficiency. It is the case of EA II where the number of necessary iterations is lower at about 100 iterations i.e., c.a. 10%. However, taking into account smaller population size in AE I, one can not in advance recommend which of the two algorithms is better.

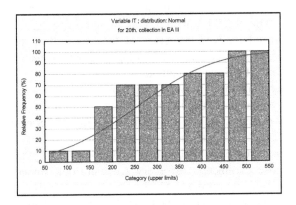

Figure 9. Histogram and cumulative frequency graph for iteration numbers at set 20 (EA III)

In EA III algorithm the exchanging crossover applied in EA II was replaced with averaging crossover. This led to the decrease in number of necessary iterations at about 40% as compared to EA II, but also the EA III population size decreased at about 20%. This made EA III much more than algorithms EA I and EA II.

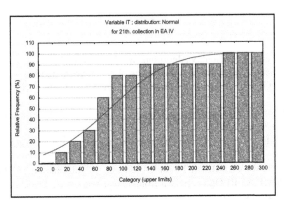

Figure 10. Histogram and cumulative frequency graph for iteration numbers at set 21(EA IV)

Algorithm EA IV required 45% less iterations than algorithm EA III. Although the difference in size of population was beneficial for EA III and was smaller at 17%, however, as it can easily be seen EA IV is the most efficient algorithm.

4.   SUMMARY AND CONCLUSIONS

This work presents the results of the research on the effectiveness of four evolutionary algorithms applied in a given mechanism trajectory optimization task. The fitness function was to minimize the differences between the given and realized trajectory of the mechanism. In order to compare the algorithms effectiveness a computing experiment has been devised applying the Latin Squares method.

The two criteria for the AE evaluation were: the number of iterations necessary to achieve the required fitness function value of the mechanism, and the probability of acquiring that value with a given number of iterations. Four algorithms were tested in terms of the influence on the two criteria of such parameters as: crossover probability, mutation and population size. Optimal sets of those parameters for every AE were determined, and the results were statistically interpreted. All that facilitates the user to realize, which of the tested algorithms is most suitable for a particular application.

The approach presented here may be helpful in making objective comparison of evolutionary algorithms and in defining their optimal settings. It makes also possible to relate the number of iterations with expected probability of achievement of the required fitness function value.

The above conclusions can be probably extended for other optimisation problems of similar feature.

References

[1] Arabas J., Lectures with Evolutionary Algorithm, WNT, Warsaw, 2001
[2] Goldberg D.E., Genetic Algorithms in Applications, WNT, Warsaw 1989.
[3] Michalewicz Z., Genetic Algorithms+Data Structures=Evolution Programs, WNT, Warsaw 1999.
[4] Miler S., Theory of Mechanisms and Machines – Synthesis of Mechanical Systems, WPW, Wrocław 1979. (In Polish)
[5] Morecki A., Oderfeld J., Theory of Mechanisms and Machines, PWN, Warszawa, 1987. (In Polish)
[6] Rohatyński R., Frąckowiak Ł, Klekiel T., An Evolutionary Algorithm in the Optimal Four-bar Linkage Synthesis, XVIII Ogólnopolska Konferencja Naukowo - Dydaktyczna TMM, Lądek Zdrój, 2002. (In Polish)

MINIMUM HEAT LOSSES SUBJECTED TO STIFFNESS CONSTRAINTS:WINDOW FRAME OPTIMIZATION

R. A. Białecki[1] and M. Król[2]

[1]*Institute of Thermal Technology, Silesian University of Technology, Konarskiego 22, 44-100 Gliwice, Poland*
[2]*Department of Heat Supply, Ventilation and Dust Removal Technology, Silesian University of Technology, Konarskiego 20, 44-100 Gliwice, Poland*

Abstract: Genetic algorithm has been used to find an optimal design of the movable and fixed window framework taking into account their thermal interaction with the glass pane and the wall. The objective function has been defined as a minimum heat losses subjected to a constraint of constant stiffness and amount of steel. To accelerate the computations parallel processing has been implemented.

Keywords: genetic algorithms, window frame, heat losses, parallel processing

1.   INTRODUCTION

The main driving forces of reducing the heat losses from buildings are the increasing energy costs. Typically, about 30% of heat is lost through windows. Standard windows consist of double glazed panes and wooden, plastic or metal frames. Window frames have smaller surface area than the window panes, thus for a longer time the optimal thermal design of these elements has been of secondary importance. At the current level of glazing and walls insulation the question of heat losses from window frames becomes important.

The present paper deals with an optimal design of a plastic window frame. To increase the insulating properties of the plastic frames, air

T. Burczyński and A. Osyczka (eds),
IUTAM Symposium on Evolutionary Methods in Mechanics, 23–32.
© *2004 Kluwer Academic Publishers.*

chambers are introduced. However, such profiles do not have the required stiffness. The necessary stiffness is achieved by inserting metal profiles in the frame. The presence of a highly conducting metal increases the heat losses. Obviously, the shape of stiffener has a greatest influence on the heat losses through the frame.

The determination of the optimal geometry of the frame in the sense of minimum heat losses subjected to the constraints has been accomplished using a standard genetic algorithm [1]. Genetic algorithms, whose principle mimics the natural selection process, offer an elegant way of circumventing the disadvantages of the standard optimization techniques. The algorithms do not require calculation of sensitivity coefficients and can readily be employed to problems with varying topology. Robustness of these procedure in the presence of local minima is their another important advantage. On the other hand, the computing times of genetic algorithms are much longer than in the case of standard nonlinear programming. Only the recent drastic drop of the computing costs along with the parallel computing options made algorithms competitive with the standard optimization techniques.

The first step of the study was the optimization of a simplified geometry. It was a fixed frame with neglected influence of heat transfer to the pane and the wall. Some examples were already discussed in our previous work [2,3]. Depending on the imposed constraints the optimization lead to a reduction of the heat losses by 10% to 30%.

In the present study an optimization of a more realistic configuration has been undertaken. The computational domain encompassed the movable and fixed window framework with thermal interaction with the glass pane and the wall taken into account. In this case the shape is quite complex and fine discretization is required to ensure sufficient accuracy. As a result the solution of a boundary value problem is time consuming. The heat transfer solver used in this study based on the Boundary Element Method (BEM) needed about 18 minutes to calculate the heat losses (Athlon 1GHz, in-core solver). The grid (limited to the boundary) consisted of 1797 nodes and 821 elements Additionally, as two stiffeners need to be considered, the number of design parameters is significant. This is turn makes the number of shapes that need to be considered high. The entire optimization process is therefore numerically intensive. The estimated processing time is in this case of the order of two months. To accelerate the computations parallel processing has to be implemented.

2.   FORMULATION OF THE PROBLEM

2.1   Heat transfer

A 2D problem steady state heat transfer is considered. The frame with a portion of wall and pane consists of seven materials: PVC, air, steel, ceramic airbrick, glass, aluminum and rubber. For the air in the window pane the equivalent heat conductivity has been calculating according to the equation of Shewen [4]. The values taken in the calculations are shown in Table 1.

Table 1. Material properties used in the calculations.

material	heat conduc. W/m K	material	heat conduc. W/m K
steel	28,0	PVC	0,163
brick	0,30	aluminum	164,0
glass	0,80	air in the chamber	0,023
rubber	0,20	air in the win. pane	0,040

Prescribed boundary conditions are depicted in Figure 1. On the portions of the contour exposed to the environment and in contact with the air in the room Robin's boundary conditions are prescribed.

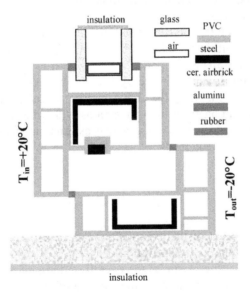

Figure 1. Geometry and prescribed boundary conditions for the window frame with portion of the pane and the wall. Wall and pane not to scale.

For the temperature differences and geometrical dimensions occurring in the problem, both natural convection and radiation are of minor importance in the air filled enclosures in the frame. Thus, it is assumed that the heat in the cavities is transferred solely by conduction.

The values of the outdoor and indoor temperatures were set to +20° and -20° which is in agreement with Polish standards [5,6]. The values of the indoor and outdoor film coefficients 23 and 8 W/m^2K have been taken from another Polish standard [7]. Heat transfer through the remaining portions of the external surface of the frame has been neglected. On the interfaces between different material ideal thermal contact, ie. continuity of both temperature and heat flux has been assumed. The geometry of the numerical examples is a simplified version of a real frame taken from [8].

2.2   Formulation of the optimization problem

The objective of the optimization is to minimize the heat losses subjected to several constraints [2,3]:
a) stiffness should be greater then a prescribed value
b) amount of steel used should not exceed some specified value
c) the stiffeners should be contained within air cavities
d) outer contour of the frame should not change
e) minimum thickness of the plastic walls should be greater than some prescribed values. Different values of minimum wall thickness for internal and external walls were applied.
f) geometry of the frame is approximated by a set of line segments
It is assumed that the element of the frame can be modeled as a beam. Additional stiffness resulting from the connections with other elements of the frame is neglected, which is a conservative assumption. Standard 1D beam equation used in the study reads

$$EI_{yy} \frac{d^4u}{dx^4} = 0$$

where u is the deflection of the axis of the beam, E and I_{yy} are Young modulus and moment of inertia, respectively.

As the contribution of the plastic to the overall stiffness of the frame is negligible, the measure of the stiffness is the moment of inertia of the metal insert with respect to a vertical (y) axis passing through the center of gravity.

The design variables are contractions, expansions and translations of the air cavities, and deformations of the steel stiffeners. The location of the characteristic points of the boundary, ie. the corner points of the air cavities and the stiffener is expressed in terms of decision variables defined as the coordinates of some control points. In the developed algorithm, the coordinates of the characteristic points are defined as an arbitrary linear combination of the control points. This approach offers significant flexibility in defining the admissible variation of the geometry.

3.   NUMERICAL TECHNIQUE

3.1   Solution of the heat conduction problem

The heat losses from the frame have been computed using BETTI, a boundary element code [9]. The first step of the boundary element method (BEM) is a transformation of the original boundary value problem in homogeneous domain into an equivalent integral equation of a form [10].

$$c(\mathbf{p})T(\mathbf{p}) = \int_C [q(\mathbf{r})T^*(\mathbf{p},\mathbf{r}) \square T(\mathbf{r})q^*(\mathbf{p},\mathbf{r})]dC(\mathbf{r}) \tag{1}$$

Where $\mathbf{r}$, $\mathbf{p}$ are vector coordinates of the current and observation points, respectively. T is the temperature and q the associated heat flux $q=-k\square T$ $\mathbf{n}$ with k standing for the heat conductivity and $\mathbf{n}$ being the outward drawn unit normal vector of the contour. T^* is the fundamental solution of the Laplace equation and $q^*=-k\square T^*$ $\mathbf{n}$. $c(\mathbf{p})$ is fraction of the angle subtended in the computational domain. The vertex of the angle is at point $\mathbf{p}$.

The next step is the discretization of eq. (1). This is done by approximating the geometry, temperature and normal flux using locally based shape functions. The final set of equations is then generated by nodal collocation. The result reads

$$\mathbf{H}^i\mathbf{T}^i + \mathbf{G}^i\mathbf{q}^i = 0 \tag{2}$$

Where $\mathbf{H}$ and $\mathbf{G}$ are the influence matrices and vectors $\mathbf{T}$ and $\mathbf{q}$ gather values of temperatures and heat fluxes at nodes located on the boundary. Superscript i refers to the number of the subregion.

The procedure is repeated in all subregions and the sets of linear equations corresponding to subregions are linked by enforcing the continuity of temperatures and heat fluxes on the interface between adjacent subregions. The details of the BEM technique are available in standard references [10].

3.2   Checking the stiffness

Checking the satisfaction of the constraints requires calculation of surface area, coordinates of the mass center and the moment of inertia. All these quantities are expressed in terms of surface integrals

$$A = \int_A dA(\mathbf{r}), \quad x_0 = \int_A x dA(\mathbf{r})/A, \quad I_{yy} = \int_A (x \square x_0) dA(\mathbf{r}) \tag{3}$$

where A is the surface area, x_0 is the x coordinate of the mass center, $I_{\{yy}$ moment of inertia with respect to the y axis passing through the mass center.

The evaluation of these surface integrals can be significantly simplified by converting them into contour integrals. This has been accomplished making use of the Stokes theorem.

$$\oint_C \vec{w} d\vec{C}(\mathbf{r}) = \int_A rot\vec{w} d\vec{A}(\mathbf{r}) \tag{4}$$

where $\vec{w}$ is an arbitrary vector, C is a contour of the surface A. As the surface of integration lies in the xy plane the normal infinitesimal surface vector is defined as $\vec{A} = \{0, 0, dxdy\}$ and the vector tangential to the contour line has a form of . $\vec{C} = \{dx, dy, 0\}$. Denoting by w_A, w_y, w_I vectors used to calculate the surface area, center of gravity and moment of inertia, respectively, their rotations are defined as

$$rot\vec{w}_A = \{0,0,1\}, \quad rot\vec{w}_x = \{0,0,x\}, \quad rot\vec{w}_I = \{0,0,(x \square x_0)^2\} \tag{5}$$

It can be readily proved that the vectors $\vec{w}$ should be defined as

$$\vec{w}_A = \{0, x, 1\}, \quad \vec{w}_x = \{\square xy, 0, 0\}, \quad \vec{w}_I = \{\square y(x \square x_0)^2, 0, 0\} \tag{6}$$

Making use of the parametric representation of the line segments of the contour, the original surface areas can be converted into boundary integrals. This simplifies the calculations of the constraints and is in line with BEM, the numerical technique used in the study, where only boundary integration needs to be carried out.

3.3   Genetic algorithm

The evaluation of the optimal geometry of the frame in the sense of minimum heat losses subjected to the constrains defined in the previous section have been accomplished using a standard genetic algorithm.

The procedure starts with a creation of an initial population consisting of N_G members. The fitness function is expressed in terms of heat losses Q_L by an relationship fitness $= (Q_L)^{-p}$ where p is a user defined constant.

In the next steps new generations are created. The number of individuals in a generation does not change throughout the iterative process. The new generation is generated in three stages: selection, mutation and mating.

The main features of the implemented version of the algorithm are given in previous works [2,3].

3.4   Influence of the parameters of the genetic algorithm

The efficiency and accuracy of the genetic algorithm depend on the values of a set of tuning parameters
- □   number of individuals in the generation N_g
- □   number of populations N_p
- □   probability of mutation P_m
- □   probability of crossover P_c
- □   power used in the definition of the fitness function p.

In the present work a proper selection of those parameters are very important. Finding a plausible set of these parameters has been accomplished by solving a simpler problem [2,3]. All the tests were made for a configuration consisting of the fixed frame insulated from the movable frame as well as the wall and the glass pane. As the computing times were short, several test could be run in order to find the best set of parameters controlling the genetic algorithm. It is clear, that this set is not optimal for the problem under consideration, but due to the similar nature of the both problems, their tuning parameters should not differ too much

The values of the tuning parameters applied when solving the problem were $N=16$, $P_m = 0.15$, $P_c = 0.5$, $p=1$. The number of populations has been determined interactively by analyzing the values of the fitness of the best individual in subsequent populations. The results presented in the papers correspond to $N_p = 500$.

4.   SELECTION OF PROPER DEGREES OF FREEDOM

It is fairly difficult to decide which elements of the fixed and movable frame should change their position to minimize the heat losses. The choice need to be made in a heuristic way, as there is no methodology of selecting the best set of control variables. As the computing times are too long to make numerical experiments, common sense need to be used to identify the variables controlling the nonlinear programming procedure.

The final configuration of the degrees of freedom of fixed and movable frame can be seen in Figure 2.

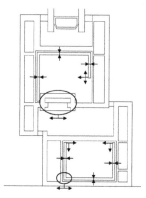

Figure 2. Design parameters used in the frame.

5.   PARALLEL PROCESSING

Our computer cluster consists of nine PCs (Athlon 1GHz). An important constraint in using the cluster was that during the computations, the computers should be available to other users. Using standard parallel processing software (PVM, MPI) could in this case be cumbersome, as during the computations the workload from other tasks could dynamically change. Therefore, we decided to develop own parallel computation software, to take care of the specific needs.

Here are some features of the developed package. The source code of the genetic algorithm does not undergo any changes and resides on the server playing the role of the supervisor of the entire process. The supervisor module generates the chromosomes of the next population (input files for BETTI), distributes the workload and controls the state of the current computations on each workstation. Checking the presence of specific files within the shared disk space controls the progress of computations. The evaluation of the fitness function ie, the solution of heat conduction problem using BEM is carried out at a cluster of workstations. The communication between server and workstations is supported by disk mapping mechanism.

In the course of computations the software is able to reschedule tasks allotted to a given workstation in the case the computations are stalling. Also, a previously excluded workstation can automatically be assigned new tasks. Thus, the software is insensitive to failures of the hardware and software.

6.   RESULTS

The combination of BEM and genetic algorithms proved to be an efficient tool to solve this fairy complex shape optimization problems. The simple remeshing, limited only to the boundary, characteristic to the BEM is its great advantage in this context. Though the calculations are time consuming, the paralellization mitigates this problem significantly.

The calculations carried out by the authors show the possibility of noticeable reduce the heat losses from a window frame. This can be achieved by simple modification of the geometry of the plastic frame and the steel stiffeners. Figures 3 and 4 show the initial and resulting geometry of the frame. Glass panes and wall are not shown in the figure.

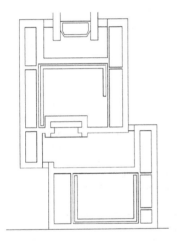 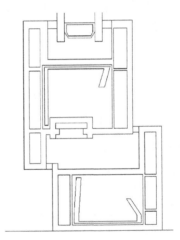

Figure 3. Starting configuration of the frame.   *Figure 4.* Resulting geometry of the frame after optimization.

The final result of the optimization was a reduction of the heat losses from 6.74 W/m to 5.97 W/m. It means the reduction of about 12%. It should be stressed that the results obtained in this study are obtained assuming several simplifications concerning the mechanisms of heat transfer in the frame. To yield a better estimation concerning the reduction of the heat losses, the heat transfer problems in the initial and final configurations will be recomputed using a CFD code taking into account natural convection and radiation within the cavities.

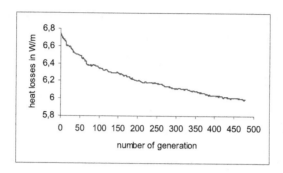

Figure 5. Reduction of the heat losses in the course of iteration.

ACKNOWLEDGEMENT

This research was carried out in the framework of the Polish Committee for Scientific Research under the grant 4 T10B 023 23.

References

[1] D. E. Goldberg; *Genetic Algorithms in Search, Optimization and Machine Learning*, Addison-Wesley, Reading, MA, 1989
[2] M. Król, R.A. Białecki; *Optimization of the window frame by genetic algorithm*, Proc. AI-MECH 2001, International Seminar on Methods of Artificial Intelligence in Mechanics and Mechanical Engineering, T. Burczyński and W. Cholewa eds, Gliwice, (2001) pp 103-108.
[3] M. Król, R. A. Białecki; Optimization of the window frame by BEM and genetic algorithm; International Journal for Numerical Methods in Heat and Fluid Flow, Accepted for publication.
[4] W. M. Rohsenow, J. P. Hartnett, Y. I. Cho - *Handbook of Heat Transfer*, McGraw-Hill Book Company, United States,
[5] Polish Standard PN-82/B-02402 *Heating; Temperature for heated rooms in building*,
[6] Polish Standard PN-82/B-02403 *Heating; Exterior calculated temperatures*,
[7] Polish Standard PN-EN ISO 6946 *Building components and building elements. Heat resistance and thermal transmittance – calculations methods,*
[8] Technical approval ITB AT-15-2045/98. *Windows and balcony doors of the KBE AD and KBE MD systems of plastified PVC sections.* Institute of Building Technology, Warsaw,
[9] R. A. Białecki, G. Kuhn; *Upgrading BETTI: introducing nonlinear material, heat radiation and multiple right hand sides options*, Rept.No.019 2 0199894 2 9 under contract with Mercedes Benz A. G., Lehrstuhl fur Technische Mechanik, Universitat Erlangen Nurnberg, Erlangen, Germany, 1993
[10]C. A. Brebbia, J. C. F. Telles, L. C. Wrobel; *Boundary element techniques, theory and applications in engineering.* New York, Springer Verlag, 1984,

EVOLUTIONARY COMPUTATION IN INVERSE PROBLEMS

T.Burczyński[1,2], E. Majchrzak[1], W. Kuś[1], P. Orantek[1], M. Dziewoński[1]

[1]*Department for Strength of Materials and Computational Mechanics, Silesian University of Technology, Konarskiego 18a, 44-100 Gliwice, Poland.*
[2]*Institute of Computer Modelling, Cracow University of Technology, Warszawska 24, 31-155 Cracow, Poland. Tadeusz.Burczynski@polsl.pl*

Abstract: Evolutionary computations in identification of multiple material defects (voids and cracks) in mechanical systems and identification of shape and position of a tumor region in the biological tissue domain are presented. The identification belongs to inverse problems and is treated here as an output (measurement) error minimization, which is solved using numerical optimization methods. The output error is defined in the form of a functional of boundary displacements or temperature fields. An evolutionary algorithm is employed to minimize of the functional. Numerical tests of internal defects identification and some anomalies in the tissue are presented.

Keywords: evolutionary algorithms, inverse problems, identification of defects, detection of a tumor, the finite element method, the boundary element method

1.   INTRODUCTION

Most of the catastrophic failure of mechanical structures were caused by the appearance of material defects. There are several non-destructive methods, applied in identification of such defects but only a few of them are able to find internal defects, which in some cases are very hardly detectable.

Evolutionary algorithms can be very useful in solving of inverse problems. They were used in identification problems in [1],[2],[3],[4], [11].

T. Burczyński and A. Osyczka (eds),
IUTAM Symposium on Evolutionary Methods in Mechanics, 33–46.
© *2004 Kluwer Academic Publishers.*

The goal of the proposed work is to develop and examine a solution technique for non-destructive identification of multiple internal defects (cracks and voids) in mechanical systems being under dynamical loads and tumor identification in biological systems having measured temperature field. This technique is based on minimization approach performed by the evolutionary algorithm and using the boundary element method and finite element method to solve direct problems.

Evolutionary algorithms were used in identification problems in [2], [3], [4], [7] and [8]. The paper deals with the identification of multiple internal defects in mechanical systems being under dynamical loads and tumor identification having measured temeperature field on the skin surface. In order to solve the defect identification problem the evolutionary approach is proposed.

2.   FORMULATION OF DEFECT IDENTIFICATION

Consider a bounded body B with an external boundary S, containing an internal defect in the form of a void V of the boundary Γ (Fig. 1a) or a crack with the crack surface Γ (Fig. 1b). Let Ω denote the actual body (i.e. containing the defect): $\Omega = B\backslash V$ or $\Omega = B\backslash\Gamma$ and $\partial\Omega = S \cup \Gamma$. The displacement u, strain ε and stress σ are related by well-known field equations of linear elastodynamics in the time domain:

$$\text{div}\,\sigma - \rho\ddot{u} = 0$$
$$\sigma = C : \varepsilon \qquad\qquad (1)$$
$$\varepsilon = \frac{1}{2}\left(\nabla u + \nabla^{T} u\right)$$

where ρ - a material density, C – a fourth-order elasticity tensor.

Eqs (1) are completed with boundary and initial conditions. The given traction $\bar{p}$ is imposed on a part of the external boundary S, while on the rest of S the displacement $\bar{u}$ is known. The boundary Γ is traction-free and the initial rest is assumed. The traction vector $p = \sigma n$ is defined in terms of the outward unit normal $\mathbf{n}$ to boundary S. In the crack case the displacement u is allowed to jump across Γ; $[\![u]\!] = u^{+} - u^{-} \neq 0$.

If the body undergoes free vibration, the governing equation is described as follows:

$$\text{div}\,\sigma + \omega^{2}\rho u = 0 \qquad\qquad (2)$$

where ω denotes a circular eigenfrequency of the body.

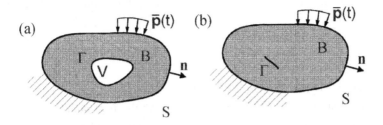

Figure 1. A body with an internal defect: a) void, b) crack

Consider the problem of finding the shape and position of an internal defect using elastodynamics experimental (or, for instance presented in this paper, simulated) data. The lack of information about V and $\square$ is compensated by some knowledge about u on S or $\square$ (redundant boundary data). The usual approach for finding $\square$ is the minimization of some distance J between u or $\square$ computed for an arbitrary internal defect and $\hat{u}$ or $\square$ measured or simulated (computed for the actual defects), e.g.:

$$J = w_1 J_1 + w_2 J_2 \tag{3}$$

where w_1 and w_2 are weight coefficients, J_1 and J_2 are defined as follows:

$$J_1 = \frac{1}{2}\sum_{i=1}^{N}\left(\hat{\square}_i \square \square_i\right)^2 \tag{4}$$

$$J_2 = \int_0^T \int_{S_m} \square\left[u(x,t)\right] dSdt = \frac{1}{2}\int_0^T \int_S \left[\hat{u}(x,t)\square u(x,t)\right]^2 dSdt \tag{5}$$

where $\square_i$ indicates i-th circular eigenfrequency of the body, $u(x,t)$ is a displacement vector of the point x on the boundary S at time t.
In order to evaluate (4) and (5) the boundary element method was used.

3.   FORMULATION OF TUMOR IDENTIFICATION

The body surface temperature of biological systems depends on the local metabolism, the blood circulation and heat exchange between the skin and the environment. These parameters can change if there are some anomalies in the body like a tumor [7],[8],[9]. An application of evolutionary algorithms in position and shape identification of these anomalies based on known surface temperature values is considered.

The bioheat transfer in the tissue is described by the Pennes equation in the form [7],[8]:

$$\lambda \nabla^2 T(\mathbf{x}) + Q_{perf} + Q_m = 0, \qquad \mathbf{x} \in \Omega \tag{6}$$

where λ [W/mK] is the thermal conductivity of the tissue, Q_{perf} [W/ m^3] is the perfusion heat source, Q_m [W/m^3] is the metabolic heat source, T($\mathbf{x}$) is the tissue temperature at point $\mathbf{x}$=(x,y). This equation is supplemented by the boundary conditions – Figure 2.

$$q(\mathbf{x}) = -\lambda \frac{\partial T(\mathbf{x})}{\partial n} = 0, \ \mathbf{x} \in \Gamma_1$$

$$q(\mathbf{x}) = 0, \qquad \mathbf{x} \in \Gamma_2$$
$$T(\mathbf{x}) = T_a, \qquad \mathbf{x} \in \Gamma_3 \tag{7}$$
$$q(\mathbf{x}) = 0, \qquad \mathbf{x} \in \Gamma_4$$

where q($\mathbf{x}$) is the boundary heat flux, $\partial T / \partial n$ denotes the normal derivative at the boundary point considered, T_a is the arterial blood temperature [2]. It should be pointed out that the adiabatic condition on the skin surface Γ_1 enables avoiding the influence from surrounding environment, which means that the skin is covered with an insulating material. The value of the heat source $Q=Q_{perf}+Q_m$ is different for the healthy tissue and the tumor region, this means

$$Q = \begin{cases} Q_1 & \text{- healthy tissue} \\ Q_2 & \text{- tumor region} \end{cases} \tag{8}$$

In order to solve the direct problem the boundary element method (BEM) and the finite element method (FEM) can be applied. In this paper the FEM is used.

The bioheat transfer in the healthy tissue and the tissue with a circular tumor is considered. The geometry and boundary conditions are shown in Figure 2. The temperature fields in the tissue for the healthy tissue and the tissue with the tumor are presented in Figure 3. The temperature distributions over the skin surface for both cases are presented in Figure 4.

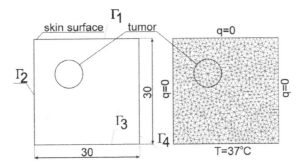

Figure 2. Tissue geometry a) before, b) after FEM discretization

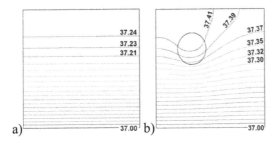

Figure 3. Temperature field: a) healthy tissue, b) tissue with tumor

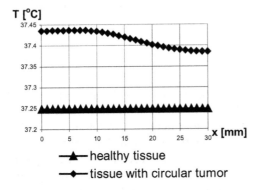

Figure 4. Skin surface temperature distribution

The shape and position of the tumor can be found by minimizing a functional formulated as a distance J between measured skin surface temperatures $\hat{T}(\mathbf{x})$ and temperatures $T(\mathbf{x})$ obtained using numerical model, e.g.

$$J = \int_{\Omega} \left(\hat{T}(\mathbf{x}) \square T(\mathbf{x}) \right)^2 d\square(\mathbf{x}) \tag{9}$$

If temperature field is measured and computed in selected sensor points $\mathbf{x}=\mathbf{x}^s$, s=1,2,..,S, then the functional (9) takes the form

$$J = \sum_{s=1}^{S} \int_{\Omega} \left(\hat{T}(\mathbf{x}) \square T(\mathbf{x}) \right)^2 \square(\mathbf{x} \square \mathbf{x}^s) d\square(\mathbf{x}) \tag{10}$$

where S is the number of sensor, $\square$ is the Dirac function.

4.   EVOLUTIONARY IDENTIFICATION METHODS

The algorithm minimizes the fitness functions (3) or (10) with respect to defect or tumor shape parameters. A vector chromosome characterizes the solution:

$$\mathbf{z} = \{z_1, z_2 \ldots z_i \ldots z_n\} \tag{11}$$

where z_i are genes which parameterize the defect or the tumor.

The genes are real numbers on which constraints are imposed in the form:

$$z_{iL} \square z_i \square z_{iR} \; ; \; i=1,2,\ldots n \tag{12}$$

The evolutionary algorithm starts with an initial generation. This generation consists of N chromosomes generated in a random way. Every gene is taken from the feasible domain. Evolutionary operators: mutation and crossover modify the initial generation. The next stage is an evaluation of the fitness function for every chromosome and the selection is employed. The selection is performed in the form of the ranking selection or the tournament selection [10]. The next generation is created and operators work for this generation and the process is repeated. The algorithm is stopped if the chromosome, for which the value of the fitness function is zero, has been found. An effectiveness of the evolutionary algorithm depends on its operators, which can be defined in a different way. The flowchart of the evolutionary algorithm is presented in Fig. 5.

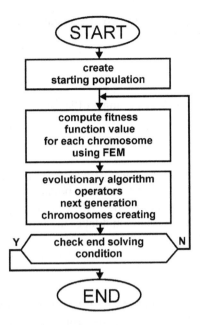

Figure 5. Flowchart of evolutionary algorithm

The crossover operator swaps some chromosome of the selected parents in order to create the offspring. Simple, arithmetical and heuristic crossover operators are used.

The *simple crossover* needs two parents and produces two descendants. The simple crossover may produce the offspring outside the design space. To avoid this, a parameter $\square\square[0,1]$ is applied. For a randomly generated crossing parameter *i* it works as follows (chromosomes $\mathbf{z}_1$, $\mathbf{z}_2$ are parents p1 and p2, respectively, and chromosomes $\mathbf{z}_1', \mathbf{z}_2'$ are children d1 and d2, respectively):

$$\text{p1}: \mathbf{z}_1 = \left\{ z_1, z_2, \ldots, z_i, \ldots, z_n \right\}$$
$$\text{p2}: \mathbf{z}_2 = \left\{ e_1, e_2, \ldots, e_i, \ldots, e_n \right\} \tag{13}$$

d1:

$$\mathbf{z}_1' = \left\{ z_1, \ldots, z_i, +\square e_{i+1} + (1\,\square\square) z_{i+1}, \ldots, \square e_n + (1\,\square\square) z_n \right\} \tag{14}$$

d2:

$$\mathbf{z}_2' = \left\{ e_1, \ldots, e_i, +\square z_{i+1} + (1\,\square\square) e_{i+1}, \ldots, \square z_n + (1\,\square\square) e_n \right\} \tag{15}$$

The *arithmetical crossover* gives two descendants, which are a linear combination of two parents

$$\mathbf{z}_1' = \square\mathbf{z}_1 + (1\,\square\,\square)\mathbf{z}_2 \; ; \quad \mathbf{z}_2' = \square\mathbf{z}_2 + (1\,\square\,\square)\mathbf{z}_1 \tag{16}$$

The *heuristic crossover* produces a single offspring:

$$\mathbf{z}_1' = r(\mathbf{z}_2\,\square\,\mathbf{z}_1) + \mathbf{z}_2 \tag{17}$$

where r is a random value from the range $[0,1]$ and $J(\mathbf{z}_2)\,\square\,J(\mathbf{z}_1)$.

Four kinds of mutation operators: uniform, boundary and non-uniform mutation, are used. The chromosome before mutation has the form $\mathbf{z}_1 = \{z_1, z_2 \ldots, z_i, \ldots, z_n\}$ and after mutation it takes the form

$$\mathbf{z}_1' = \{z_1, z_2 \ldots, z_i', \ldots, z_n\} \tag{18}$$

The *uniform mutation:* children are allowed to move freely within the feasible domain and the gene z_i' takes any arbitrary value from the range $[z_{iL}, z_{iR}]$.

The *boundary mutation:* the chromosome can take only boundary values of the design space, $z_i' = z_{iL}$ or $z_i' = z_{iR}$.

The *non-uniform mutation:* This operator depends on generation number t and is employed in order to tune of the system

$$\mathbf{z}_i' = \begin{cases} z_i + \square\left(t, z_{iR}\,\square\,z_i\right) & \text{if a random digit is 0} \\ z_i\,\square\,\square\left(t, z_i\,\square\,z_{iL}\right) & \text{if a random digit is 1} \end{cases} \tag{19}$$

where the function $\square$ takes value from the range $[0, e]$.

5.   GEOMETRICAL PARAMETERIZATION OF DEFECTS AND TUMORS

The material defect is parameterized as an elliptical flaw (Fig. 6). In this case the chromosome, for the *i-th* flaw, consists of five genes

$$\mathbf{z}^i = \{z_1, z_2, z_3, z_4, z_5\} \tag{20}$$

where $z_1=x$ and $z_2=y$ are co-ordinates of the center of the flaw, $z_3=r_1$ and $z_4=r_2$ are radii of the flaw and $z_5=\square$ is an angle.

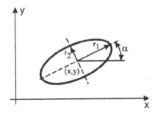

Figure 6. Parameterization of the elliptical flaw

From the elliptical flaw one can obtain special material defects as:
- circular void if $r_1=r_2=r$ and $\square=0$,
- crack if $r_2 \leq r_{min}$, where r_{min} is a prescribed admisible small value.

In the case if $r_1 \leq r_{min}$ and $r_2 \leq r_{min}$ the defect does not exist.

For the case when the body contains n defects the chromosome takes the form

$$\mathbf{z}=\{\mathbf{z}^1, \mathbf{z}^2, ..\mathbf{z}^i.,\mathbf{z}^n\} \tag{21}$$

where $\mathbf{z}^i$ is the vector which contains genes for the *i-th* elliptical flaw.

The NURBS curve can represent areas of the tissue with a tumor. The chromosome contains information about the center of the tumor (x,y) and coordinates of control points P_1-P_n. These coordinates are relative to the center of the tumor. The chromosome takes the form:

$$\mathbf{z} = \{x, y, P_{1x}, P_{1y}, P_{2x}, P_{2y}, ...\} \tag{22}$$

The sample NURBS curve with 4 control points is presented in Fig. 7.

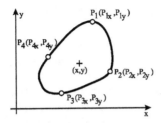

Figure 7. Parameterization of NURBS curve

6.   NUMERICAL EXAMPLES

Numerical tests of identification have been carried out for 2-D problems. The identification procedure of the defect and the tumor is based on the evolutionary algorithm and employs information about of the objective functionals (5) and (10).

6.1   Defect identification

A 2-D structure, shown in the Fig. 8 contains two internal defects. The actual parameters of an elliptic void are: $z^2=z(2)=\{50, 25, 5, 2.5, 2.5\}$, where the first two parameters are co-ordinates of the ellipse center, next - two radii of the ellipse and the last one – the angle between the x_1 axis and first radius. The actual crack parameters are: $z^1=z(1)=\{20, 30, 5, 0, 0.25\}$ and are defined as for the ellipse. The identification task is to find a number of defects and their shape having displacements $\hat{u}(x,t)$ in 33 sensor points, shown in the Fig. 8.

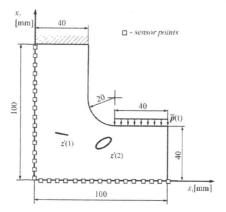

Figure 8. The 2D structure with an internal crack and void

The structure is loaded by $p(t)=p_0\sin\Box t$ ($p_0=40$ kN/m, $\Box=15708$ rad/s) in time $t\Box[0, 600\infty s]$ and has the following material properties: the Young modulus E = 0.2E12 Pa, the Poisson's ratio $\Box = 0.3$ and the density $\Box = 7800$ kg/m^3. The multiple defect identification has been solved with the assumption, that the body contains: 2 defects, 1 defect or no defect. The chromosome consists of 10 genes, where first 5 parameterize the first ellipse, and last 5 the second ellipse. When one of the genes, which is an ellipse radius is less than $r_{min}= 2$mm, the ellipse becomes a crack, when the both radii are less than r_{min} the ellipse disappears. The population contains 2000 chromosomes. The tournament method of selection was used. The solution was obtained for the case with no noise in a 100 generations and for noisy data of displacements in 120 generation. The fitness function (3) was evaluated using the boundary element method. Fig. 9 presents the best solution of the first and the last generation.

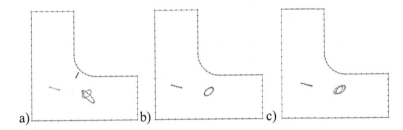

Figure 9. Results of defect identification : a) 1st generation, b) 100th generation, c) 120th generation (noisy data)

6.2 Tumor identification

The surface skin tissue temperatures called "measured" are obtained using numerical simulation for the actual shape and position of the tumor.

The domain of the biological tissue of dimensions 30x30 [mm] with a tumor is considered. The tissue properties used in numerical examples are following: thermal conductivity $\square$=0.75 [W/mK], heat generation for healthy tissue Q_1=420 [W/m^3], heat generation for tissue with a tumor Q_2=4200 [W/m^3].

The sensor points are distributed on the top boundary of the tissue as shown in the Fig. 10. The temperature equal to 37 [°C] is applied at the bottom boundary of the tissue, the rest of the boundary is isolated (heat flux equal to zero). The tumor was modelled using NURBS curves with 4 control points. Coordinates of every control point related to 'centre' of the NURBS, and position of this 'centre' are design variables. 10 design variables and 31 sensor points were used.

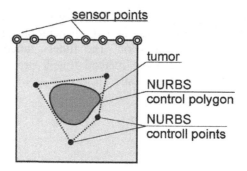

Figure 10. Geometry parametrization of the tumor by means of NURBS

The test shape of the actual tumor is presented in Fig. 11a and the best position and shape detected by the evolutionary algorithm is shown in Fig. 11b. The black area means the tumor. The skin temperature distributions for the actual and found tumor positions are shown in Fig. 12. The temperature fields for the actual and found tumor are presented in Fig. 13.

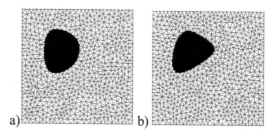

Figure 11. Finite element model of the tissue with a tumor, a) actual position of the tumor, b) the best shape and position of the tumor in the 171st generation

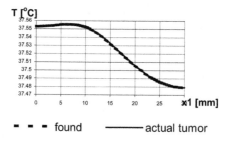

Figure 12. Skin surface temperature distribution

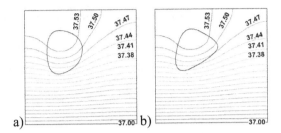

Figure 13. Temperature distribution in the tissue: a) with actual tumor, b) with the found tumor

Table 1. Comparison of design variable (gene) values

design variable (gene)	actual value	found value	error %
x	10	10.97	9.70
y	20	19.66	1.70
P_{1x}	-4	-3.95	1.25
P_{1y}	6	6.92	15.33
P_{2x}	6	5.10	15.00
P_{2y}	3	2.40	20.00
P_{3x}	5	6.36	27.20
P_{3y}	-5	-1.00	80.00
P_{4x}	-4	-6.80	70.00
P_{4y}	-8	-8.39	4.87

Numerical results presented in Table 1 show that the application of the evolutionary algorithm in position and shape identification of tissue anomalies is not very accurate due to a non-unique problem, but region near actual tumor was found.

7. CONCLUSIONS

The paper deals with the applications of evolutionary algorithms to solve inverse problems in mechanics. The main subject is the formulation of identification problems in which one should find different kinds of defects as voids or cracks in mechanical systems or anomalies in the form of cancer tissue in biological systems having measured displacement and temperature fields. Using evolutionary algorithms and BEM or FEM it is possible to detect the number of defects, kinds of defects and their shape and position.

This paper is the first attempt of the application of evolutionary computation and the finite element method to the identification of the shape and position of the tumor in the biological tissue based on measurements of the temperature filed on the skin surface. Variations of the temperature field caused by the tumor are very small and therefore the inverse problem is very delicate.

References

[1] Burczyński T. (ed.), Computational Sensitivity Analysis and Evolutionary Optimization of Systems with Geometrical Singularities, ZN KWMiMKM, Gliwice, 2002.

[2] Burczyński T., Beluch W., The identification of cracks using boundary elements and evolutionary algorithms. Engineering Analysis with Boundary Elements 25, 2001, pp. 313-322.

[3] Burczyński T., Beluch W., Długosz A., Orantek P., Nowakowski M., Evolutionary methods in inverse problems of engineering mechanics. In: Inverse Problems in Engineering Mechanics II (eds. M. Tanaka and G. S. Dulikravich), Elsevier 2000, pp. 553-562.

[4] Burczyński T., Beluch W., Długosz A., Kuś W., Nowakowski M., Orantek P., Evolutionary computation in optimization and identification. Computer Assisted Mechanics and Engineering Sciences 9, 2002, pp. 3-20.

[5] Burczyński T., Bonnet M., Fedeliński P., Nowakowski M., Sensitivityanalysis and identification of material defects in dynamical systems. Journal of Mathematical Modelling and Simulation in System Analysis SAMS, Vol. 42C4, 2002, pp. 559-574.

[6] Burdina L.M. et al: Detection of Breast Cancer with Microwave Radiometry. N2, Mammology, 1998.

[7] Jing Liu, Lisa X. Xu: Boundary information based diagnostics on the thermal states of biological bodies. Int. Jou. of Heat and Mass Transfer, vol. 43, 2000, pp. 2827-2839.

[8] Majchrzak E., Mochnacki B., Szopa R.: Numerical analysis of temperature distribution in the skin tissue with a tumor. Proc. III Symp. Biomechanics in Implantology – Ustroń 2001, Annales Academiae Medicae Silesiensis, Supl. 32, Katowice, 2001.

[9] Majchrzak E., Drozdek J.: Modelling of temperature field in the skin tissue with a tumor using boundary element method. PPAM 2001 Workshop. Ed. E. Majchrzak, B. Mochnacki, R.Wyrzykowski, Częstochowa, 2001.

[10] Michalewicz Z.: Genetic Algorithms + Data Structures = Evolutionary Programs, Springer Verlag, Berlin, 1996.

[11] Stavroulakis G. E., Antes H., Flaw identification in elastodynamics. BEM simulation with local and genetic optimization. Structural Optimization 16, 1998, pp. 162-178.

[12] Zienkiewicz O.C., Taylor R.L.: The Finite Element Method. 5th edition, Butterworth-Heinemann, Woburn, 2000.

HANG-GLIDER WING DESIGN BY GENETIC OPTIMIZATION

S. D'Angelo, M. Fantetti and E. Minisci
Departement of Aerospace Engineering, Polytechnic of Turin, 10129 Turin, Italy

Abstract: A constrained multiobjective optimization code, based on genetic algorithms, was applied to improve the aerodynamic performance of a commercial hang-glider wing in some points of the hodograph starting from the shape of an existing vehicle. In particular, the optimization process was aimed to maximize the efficiency at trim speed and to minimize the stall speed. Constraints were imposed on the desired lift and on the characteristics of the longitudinal static stability. Results, in terms of Pareto front approximation and in terms of improvements of the performance of the base wing, are presented and discussed. Even if the work has to be considered only the first step of a totally automated, numerical design approach of a hang-glider wing, obtained results demonstrate how few hours of computation can successfully substitute years of trials and errors.

Keywords: genetic algorithm, multi-objective optimization, constrained optimization, wing design.

1.   INTRODUCTION

In recent years the use of aeronautical light vehicles such as hang-gliders and para-gliders increased extensively, with significant improvements, in particular in terms of efficiency. For hang-gliders this evolution, carried out by means of a trial and error design methodology, led to the adoption of a rigid wing instead of the old flexible configuration. For rigid hang-gliders the same numerical aerodynamic design techniques used for aircraft subsonic wings can be applied.

Aerodynamic shape optimization is a complex and numerically intense task even if approximate, and relatively simple, models are adopted. Usually multimodality and roughness of the objective functions prevent the use of

47

T. Burczyński and A. Osyczka (eds),
IUTAM Symposium on Evolutionary Methods in Mechanics, 47–58.
© *2004 Kluwer Academic Publishers.*

traditional, deterministic methods.

The last decade has seen an ever increasing interest in numerical treatment of complex systems by means of methodologies based on statistical analysis and random number generation. Since no analytical information is required, these methods, allow the treatment of those systems described by multimodal and/or irregular models, at the expense of an increased computational time. Moreover a true multi-objective optimization can be performed.

Preliminary results showed that a classical optimization technique could be able to find good geometries improving the efficiency of the wing, but the same technique gets stuck on a local stationary point when is used to improve stall characteristics of the wing. As a consequence, if a bi-objective optimization is carried out by weighted sum of the objectives, changing the weights cannot prevent finding a sub-optimal Pareto front.

In this work an optimization tool based on Genetic Algorithms (GAs) [1] was applied to the three-dimensional (3D) aerodynamic multi-objective optimization of a rigid hang-glider wing.

A brief introduction on the general multi-objective and constrained problem is followed by a description of the main features of our GA. In section 3 the algorithm structure and the practical implementation are discussed. In section 4 the wing model is briefly described, and results of the optimization process are given. A final section of concluding remarks summarizes the present work and indicates future developments.

2.   PROBLEM STATEMENT

The problem that we are going to solve can be defined as follows: given N functions $f_i(\vec{x})$, the point

$$\vec{x}^* = (x_1^*, x_2^*, ..., x_n^*) \square\, F \,\square\, S \,\square\, R^n \tag{1}$$

is a solution of the multi-objective constrained maximization problem if there are no points $\vec{x} \square F$ that satisfy the logical condition

$$((\square i : f_i \,\square\, f_i^*) \,\square\, (\square i : f_i > f_i^*)) \tag{2}$$

S is the subset of R^n where optimal solutions are searched for. In most cases it can be defined by simple inequalities

$$x_j^{\min} \,\square\, x_j \,\square\, x_j^{\max} \qquad \square j = 1, ..., n \tag{3}$$

while the region F is the subset of S where the inequality constraints on the solution:

$$g_k(\vec{x}) \square 0 \qquad \square k = 1,...., m \qquad (4)$$

are satisfied.

The formulation of the maximum problem with the so called Pareto criterion not only allows the determination of the absolute maxima of the function f_i, but points out at the same time the best compromises between the different objectives of the maximization procedure.

3.   ALGORITHM STRUCTURE

In this paper a constrained, real-coded Genetic Algorithm, with elitism is used. The algorithm structure, and the adopted techniques are summarized in this paragraph.

3.1 Selection

The evaluation of the fitness parameter is based on a method for classifying the whole population, which favors the most isolated individuals in the objective function space in the first sub-class (highest dominance) of the first class (best suited with respect to problem constraints).

Once classified, the individuals are chosen for mating and mutation phases adopting the remainder stochastic sampling without replacement [1].

3.1.1   Constraint parameter

The first step in the evaluation of the fitness parameter is given by the determination of the degree of compatibility of each individual with the problem constraints. This compatibility is measured by the constraint parameter

$$cp(\vec{x}) = \left(\sum_{k=1}^{m} [s_k(\vec{x})]^p \right)^{\frac{1}{p}} \qquad (p > 0) \qquad (5)$$

where

$$s_k(\vec{x}) = \max\left(\frac{g_k(\vec{x})}{g_k^{\max}}, 0\right)$$ (6)

with $g_k^{\max} > 0$. In our application $p = 1$ is always used.

The value of the different constraints is scaled so as to give the same weight to all the constraints in the evaluation of the constraint parameter. Once the cp is evaluated for all the individuals, the population is divided in a predetermined number of classes, $1 + n_{class}$. The individuals that satisfy all the constraints ($cp = 0$) are in the first class. The remainder of the population is divided equally in the remainder groups, the second class being formed by those with the lower values of the constraint parameter and the last one by those with the higher values. In this way there is no need for *a priori* assignation of class boundaries.

3.1.2   Pareto criterion

The Pareto criterion is the basis for classifying the individuals of each class into sub-classes, and in the present work this is done by using the nondominated sorting methods as described in [2].

On one side the constraint parameters pushes the population toward the subset F of the search space S where all the constraints are satisfied. On the other side the dominance criterion pushes the evolution towards the Pareto front, determined by the non-dominated individuals of F.

3.1.3   Niche parameter

The last criterion for classifying the population is introduced in order to obtain a uniform distribution along the Pareto front, pushing the individuals towards the less crowded areas of the objective function space [2].

A more isolated individual is characterized by a lower value of the niche parameter and will get a higher rank among the other members of its subclass.

The choice is between penalizing those individuals with many copies or very near individuals in the objective function space, or favouring isolated individuals. In the absence of clear instructions on how assigning the niche parameter, the adopted value is that giving better results on test cases.

3.2   Matings and Mutations

The selected individuals are paired off with crossover probability, p_{cross}. A prescribed fraction of the so selected individuals undergoes the reproduction process with a mate randomly chosen among the other selected individuals, whereas the remaining part mates with individuals similar in terms of distance in variables space. This restriction of mating allows for a survival of the species because we obtain less lethal individuals [3], even if the mating with different individuals (hybridization) is essential for spanning the whole search space *S*, otherwise the evolution process can slow down, or can be stopped or simply not able to reach the entire Pareto front.

In order to prevent, or at least to reduce, the risk of losing good solutions, a simple form of elitism was introduced. The adopted technique can be described as follows: in the best constraints class, the current local front is divided in a predefined number of sections, and for each section non-dominated individuals that locally maximize the objective functions are marked. This mark guarantees that: a) crossover between two marked individuals is prevented, b) when a marked individual mates with a normal one, after the crossover the marked individual is preserved and one of the products is randomly chosen, c) mutation is not used on marked individuals.

Reproduction is based on the classical crossover with one or two cuts adapted to the real coded formulation, which acts in accordance with the following rule: once selected, two mating individuals interchange elements of $\bar{x}$ in the same way of the well known binary two-cuts crossover, but two of the elements, randomly chosen, are sampled from normal random distribution with mean equal to the old elements values and with variance imposed in order to have probability 0.8 to sample no more distant than 0.1 of the maximum range.

A mutation operator, which varies the gene sampling from a uniform random distribution centered in its previous value, is also introduced for enhancing diversification. Each parameter has probability p_{mut} being varied in accordance with the rule:

$$x_i^{new} = x_i + \left(rand \cdot 2 - 1\right) \cdot mrmut \cdot \left(x_i^{max} - x_i^{min}\right) \tag{7}$$

where *rand* is a random number sampled from an uniform probability density function between 0 and 1 ($0 \le rand < 1$), and *mrmut* is the maximum allowed variation (usually $0 < mrmut \le 0.5$).

After the crossover and the mutation routine a control function verifies whether or not the sampling procedures give elements that exceed the limits prescribed by eq. (3). When a variable fall outside the allowable range, its value is brought back to the nearest limit.

3.3   Convergence criterion

Finding an optimal convergence criterion that could be adopted for a generic optimization process, that is finding a convergence criterion that guaranties an optimal approximation of Pareto front (efficacy) and requires a number of objective function evaluations as low as possible (efficiency), is still an open question.

For this problem, in order to limit the computational efforts and, in the same time, to obtain clear instructions on how to change geometries of the base wing, the maximum generation number criterion is used.

4.   WING SHAPE OPTIMIZATION

The optimization algorithm was implemented so as to optimize the wing shape of a commercial hang-glider, with prescribed constraints.

4.1   Wing Aerodynamics and Stability

The aerodynamic characteristics of the wings are evaluated from the implementation of a modified version of the Prandtl model [4], that provides aerodynamic performance of a wing, flying at low speeds, in terms of dimensionless coefficients (lift, drag, and pitching moment) on the basis of the knowledge of airfoils characteristics. Comparison with experimental test cases data showed a good correspondence in the linear field and minor errors in the stall range.

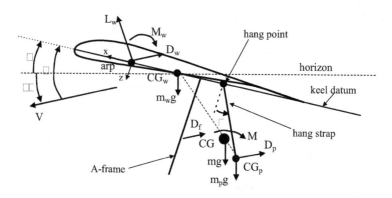

Figure 1. Longitudinal model of the hang-glider

In the present case, the aerodynamic characteristics (in terms of lift, drag, and pitch moment coefficients as a function of angle of attack and Reynolds number) of 5 sections along the semi-wingspan, were used to obtain C_{Lw}, C_{Dw} and C_{Mw} as a function of i) the angle of attack of the central section, $\square$, ii) wing geometry, and iii) airfoils characteristics. The aerodynamic reference point (*arp*) is located at 25% of the central section chord, as predicated in fig. 1, where the geometry mass and load distribution of the system are also reported.

In order to estimate stability properties, a simple model of the longitudinal characteristics of a hang-glider vehicle (wing + pilot) is adopted [5]. Under the main simplification of assuming the pilot as a point mass with fixed drag coefficient, the model gives, in particular, the value of the control angle, $\square$ that allows to trim the vehicle, and the $C_{M\square}$ coefficient, the value of which provides the required information on the longitudinal stability of vehicle.

4.2   Wing Optimization Statement

Far from being a complete and exhaustive treatment, important aspects in a hang-glider design are the efficiency at maximum speed, the efficiency at the trim speed (usually the cruise speed) and the minimum speed (or stall speed). A low minimum speed is desirable in the take-off and landing phases, while during the flight what is important is the efficiency. Preliminary results showed that if structural effects are not taken into account, geometries maximizing the efficiency at the trim speed are much similar to those ones maximizing the efficiency at the maximum speed.

In this optic, the optimization process is aimed at determining the wing shape in terms of chord and twist, sweep and dihedral angles distribution that maximizes the aerodynamic efficiency at desired trim speed and the function $16\text{-}V_s$ (where V_s is the stall speed). Obtained solutions have to meet requirements in terms of generated lift, i.e. wings have to generate aerodynamic forces allowing the glide, and in terms of stability characteristics, i.e. vehicles should flight in stable conditions.

Design variables that parameterize the wing shape are scale factors (s.f.) between some characteristics of the original, already flying, shape and the new ones explored by genetic search. Let s_a ans s_b the s.f. linked to contiguous sections a and b, respectively, then the geometric characteristic (such as chord, longitudinal position, etc.) of the explored wing is

$$c_{ew} = c_{bw}\left[\left(\frac{y_b \square y_a}{s_b \square s_a}\right)(y \square y_a) + s_b\right] \qquad (8)$$

where c_{bw} is the characteristic of the base wing, and y_a and y_b are the y positions of sections a and b respectively.

This kind of parameterization has been used in order to avoid, or at least to limit, unfeasible configurations. An a priori analysis allowed reducing the parameter number to 11. Additional genes allow the choice of free flight conditions such as incidence and speed.

Table 1. Problem variables

Variable	Meaning	Range
1	S.f. of chord of external section	$0.8 \leq x_1 \leq 1.2$
2	S.f. of longitudinal position of external section	$0.8 \leq x_2 \leq 1.2$
3	S.f. of vertical position of external section	$0.8 \leq x_3 \leq 1.2$
4	S.f. of twist angle of external section	$0.8 \leq x_4 \leq 1.2$
5	S.f. of section chord at 52% of the semi wing span	$0.8 \leq x_5 \leq 1.2$
6	S.f. of longitudinal section position at 52% of the semi wing span	$0.8 \leq x_6 \leq 1.2$
7	S.f. of vertical section position at 58% of the semi wing span	$0.8 \leq x_7 \leq 1.2$
8	S.f. of the section at 25% of the semi wing span	$0.8 \leq x_8 \leq 1.2$
9	S.f. of chord of the central section	$0.8 \leq x_9 \leq 1.2$
10	S.f. of longitudinal section position at 5% of the semi wing span	$0.8 \leq x_{10} \leq 1.2$
11	S.f. of vertical section position at 5% of the semi wing span	$0.8 \leq x_{11} \leq 1.2$
12	Geometric incidence at trim speed [deg]	$4 \leq x_{12} \leq 15$
13	Geometric incidence at minimum speed [deg]	$14 \leq x_{13} \leq 22$
14	Minimum speed [m/s]	$6 \leq x_{14} \leq 12$

The meaning of genes is given in table 1, where longitudinal and vertical section position respectively mean x and z coordinates of the quarter point of the chord section in a system centered in the *arp* (fig. 1). Allowing scale factors from 0.8 to 1.2 means allowing a variation in the range ±20% of the base geometries (if we impose every s.f. to 1, we obtain the original wing). Bounds of x_{12}, x_{13} and x_{14} are imposed in order to give sufficient margins of search.

Parameters that could affect results, but are maintained constant for simplicity, are:
- wing span, 10.4 m;
- dimension of the A-frame normal to the keel, 1.4 m;
- hang strap length, 1.1 m;

A-frame longitudinal location and hang-point are determined, by means of an iterative procedure, in order to obtain null control force at 15 m/s (trim speed) for each individual. These values are maintained constant when minimum speed conditions are evaluated, with the only consideration that distance between the two points is usually 0.1÷0.2 m. Longitudinal position of the wing center of gravity (CG_w) is updated for the new wing planform configuration by analogy with the base shape (simplified approach).

On the basis of what reported in sec. 3.1.1, the constraint functions were assigned as follows:

$$g_1(\vec{x}) = |L - L_{des}| \leq \varepsilon \qquad for \qquad V = 15 \, m/s \qquad (9)$$

$$g_2(\vec{x}) = \left| L \square L_{des} \right| \square \square \qquad for \qquad V = V_s \qquad (10)$$

$$g_3(\vec{x}) = C_{M_\square} \qquad for \qquad V = 15\,m/s \qquad (11)$$

$$g_4(\vec{x}) = C_{M_\square} \qquad for \qquad V = V_s \qquad (12)$$

where L is calculated lift and L_{des} is the desired lift, 1300 N, resulting from the sum of assumed wing and pilot weights, respectively 350 and 950 N, with $\square$ imposed as 10 N to obtained individuals with a lift error smaller than 1%. $C_{M_\square}$ must be negative to guarantee the vehicle stability.

In table 2 values of optimization algorithm parameters used for this case are shown.

Table 2. Genetic Algorithm Parameters

Parameters	Value
Degrees of Freedom (n)	14
Number of individuals (N_{ind})	100
Niche radius	0.1
Fitness scaling coefficient	1.6
Number of hybridating individuals	30
Crossover probability (p_{cross})	0.85
Mutation probability (p_{mut})	0.15
Maximum range of variation of the operator mutation (mrmut)	0.02
Constraint classes (n_{class})	3

4.3 Wing Optimization Results

Figure 2 shows the feasible solutions reached after 100 generations. 45% of the whole population satisfies the constraints. Although solutions are not uniformly spread and a refinement by classical methods could give some wing with slightly better performance, since obtained results give enough information about how base-line geometries can be changed in order to improve performance, we can consider the task accomplished. Moreover, refinement of some solutions would increase considerably the amount of the needed computation.

It can be observed that individuals are distributed in the range from 17 to 24 for the efficiency and in the range 7.4 | 8.2 m/s for the second objective function, that corresponds to a stall speed between 7.8 and 8.6 m/s.

All the solutions appear physically feasible, although solutions minimizing the stall speed with low efficiency (less than 20) are of minor practical interest, because of a high reduction of performance in terms of efficiency at the trim speed with a small improvement in terms of stall speed. Furthermore, this kind of individuals has a negative sweep angle that could be problematic in phase of construction (fig. 3). The negative sweep angle disappears when solutions on the right side of the front are considered.

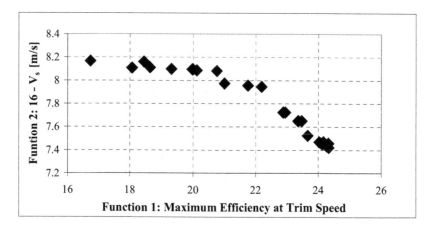

Figure 2. Feasible solutions after 100 generations

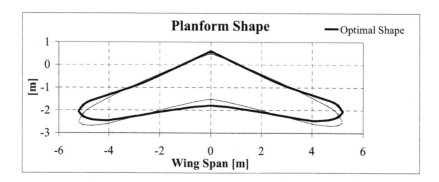

Figure 3. Plan form shape that minimize the stall speed

From a practical point of view, solutions maximizing the efficiency are more attractive, first of all because a loss in terms of stall speed results in a much better compromise with extremely higher efficiency at trim speed. Moreover, as it is evident in figure 4, where the plan form shape of the individual maximizing the efficiency at trim speed is represented, resulting wing shapes are similar to those of modern rigid wing hang-gliders, for which manufacturers claim to have 19|20 of efficiency at trim speed.

It must be noted that the validity of the proposed configuration is limited inasmuch as the performance analysis does not take into consideration the effect configuration on the hang-glider structure and only low speed regime is considered (stall speed and V = 15 m/s).

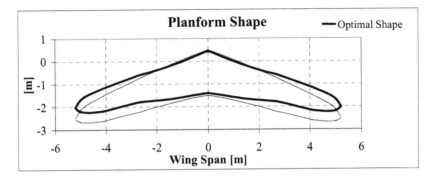

Figure 4. Plan form shape that maximize Efficiency at trim speed

5.   CONCLUSIONS

The multi-objective optimization of a rigid hang-glider wing has been performed using a real coded genetic algorithm. The optimization process was aimed to improve some aerodynamic characteristics of an existing hang-glider wing, such as the efficiency (maximization) at the trim speed and the stall speed (minimization), with constraints on the generated lift and on the longitudinal stability of the whole vehicle.

Even if the work is far from being a complete design process, it demonstrates that recent evolutionary algorithms can be suitable tools for the optimization of a hang-glider wing. In the reported case an old wing shape has been improved obtaining performance comparable to those of modern rigid wing after only few hours of computational time.

One of the main limitations of obtained results is the absence of a structural model into the formulation of the optimization problem. In this optic future works will be aimed at the implementation of a more complex and complete formulation that includes structural analysis (even simplified) as well as more accurate stability analysis (longitudinal and lateral-directional).

Moreover, the technique of niching is borrowed from a first generation Pareto technique and its limits are denounced in literature and highlighted by achieved results. A solution of this problem could be the adoption of different techniques, such as crowding and elitism techniques recently proposed in [6].

Furthermore, at this point, how and how much the variables of the problem are interconnected is not perfectly known. If the level of interconnection were high, a normal genetic algorithm would give only sub-optimal solutions (even if better of that obtained by classical methods). In this case a more efficacious approach could be by means of new different evolutionary methods, such as Estimation of Distribution Algorithms [7].

References

[1] D.E. Goldberg, Genetic Algorithms in Search, Optimization & Machine Learning, Addison Wesley, 1988.

[2] N. Srinivas and K. Deb, "Multiobjective function optimization using nondominated sorting genetic algorithms", Evolutionary Computation Journal, Vol.2, No.3, pp. 221-248, 1995.

[3] C.M. Fonseca and P.J. Fleming, "Multiobjective Genetic Algorithms Made Easy: Selection, Sharing and Mating Restriction", in Genetic Algorithms in Engineering Systems: Innovations and Applications, 1995.

[4] C. Ligorio, "Aerodinamica e Meccanica del Volo di Velivoli Ultraleggeri", Turin Polytechnic, Doctoral Thesis, 1994.

[5] M.V. Cook, "The theory of the longitudinal static stability of the hang-glider", The Aeronautical Journal, Vol.98, No.978, pp. 292-304, 1994.

[6] K. Deb, S. Agrawal, A. Pratab and T. Meyarivan, "A Fast Elitist Non-Dominated Sorting Genetic Algorithm for Multi-Objective Optimization: NSGA-II", KanGAL report 200001, Indian Institute of Technology, Kanpur, India, 2000.

[7] H. Mühlenbein, "From recombination of genes to the estimation of distributions II. Continuous parameters", Parallel Problem Solving from Nature, eds. Voigt, H.-M and Ebeling, W. and Rechenberg, I. and Schwefel, H.-P., LNCS 1141, Springer:Berlin, pp:188-197, 1996

AN ERROR FUNCTION FOR OPTIMUM DIMENSIONAL SYNTHESIS OF MECHANISMS USING GENETIC ALGORITHMS

Igor F. de Bustos, Josu Aguirrebeitia, Vicente Gomez-Garay and Rafael Aviles
Dpt. of Mech. Eng. E.S.I, Alda de Urquijo S/N, 48013 Bilbao, Vizcaya, Spain

Abstract: In this paper an error function for optimum dimensional synthesis of mechanisms using genetic algorithms is presented. This function is general for its use with any mechanism with R and P joints. This minimization function is established as the quadratic addition of the minimum linear or angular distance to the synthesis positions which can be achieved by the mechanism. To get this distance, a non-linear function must be iteratively solved, leading to a high computational cost function. In order to reduce the computational cost, variable complexity functions techniques can be applied. Solutions obtained with this genetic algorithm are a good starting point for further numerical optimization methods.

Keywords: multibody synthesis, genetic algorithms, optimization, mechanism, dimensional synthesis.

1.   INTRODUCTION

In the development of the kinematic synthesis of mechanisms, two different steps are to be addressed:
– Topology synthesis: it deals with the selection of the types of elements which are to be used in the mechanism and its configuration.
– Dimensional synthesis.
Dimensional synthesis of mechanisms deals with the problem of defining the dimensions of the elements of a mechanism in order to verify some restrictions. These restrictions can be summarized in the following points:

59

T. Burczyński and A. Osyczka (eds),
IUTAM Symposium on Evolutionary Methods in Mechanics, 59–68.
© *2004 Kluwer Academic Publishers.*

- Path generation: A point of the mechanism generates a path defined by some control points.
- Rigid body guidance: The mechanism is able to guide a solid attached in some way to it.
- Function generation: The position of an element is a given function of the position of an input link.
- Mixed: the combination of one or more of the previous ones.
  This problem has been approached in several ways.
- Graphical methods: Applicable to four-bar mechanisms with a low number of synthesis positions.
- Analytical Methods: Only applicable to simple mechanisms with a limited number of restrictions that must all be verified.
- Numerical methods: these methods are quite useful to improve a previously dimensioned mechanism in such a way that the verification of the imposed restrictions can be optimized. These lead to a good solution, provided that they start from a good initial mechanism. Restrictions are minimized, so that an unlimited number of restrictions can be introduced [1,2,3].
- Other methods: Some other methods to solve this problem have been proposed. Some of them for specific mechanisms, and others based on mechanism path databases[4,5] and selection, or neural networks [6].

The work presented in this paper tries to solve the above problem by using a new approach, taking advantage of the use of genetic algorithms. These are chosen because of the beneficial characteristics that they show:

- They work with a population of solutions. So a good exploration of the synthesis space can be made and among the solutions obtained, the user can choose one taking into account criteria not included in the error function.
- Synthesis space can be easily defined. Therefore, keypoints as fixed points can be limited to a space defined by the geometry of a real frame, for example.
- Possibility of hybridation: the introduction a numerical method in the genetic algorithm can be considered.

Mechanism synthesis is a complicated problem, thus leading to a high computational cost. The use of a variable complexity function technique can effectively reduce this computational cost making genetic algorithms an affordable tool for the effective dimensional synthesis of mechanisms.

2. PROBLEM FORMULATION

Previous work made by this group was focused on the use of numerical methods for both the kinematic analysis (speed, acceleration, and position problems including deformed position problems) and the synthesis. This work [1,2] led to a compact formulation suitable for solving the whole kinematic problem with a unified formulation. It also became a useful tool for mechanism synthesis, solving any of the possible problems, provided that a good starting point was given. This formulation showed strong stability and fast convergence with any kind of mechanism with R-joints, but also showed some convergence problems with P-joints. The formulation of the problem is quite simple, it relies upon the minimization of the deformation energy of the mechanism when its elements are assumed to be linear-elastic deformable with finite stiffness.

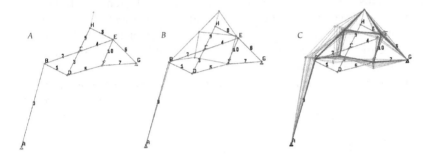

Figure 1. Deformed position problem. A: Initial position B: deformed position C: deformed position & iterations

For this deformable mechanism, when a restriction is imposed, the minimization leads to the initial position if the restrictions are obtainable without deformation, or to the deformed position if restrictions are not reachable without deformation. Therefore the function to minimize is (1).

$$\square(\{x\}) = \frac{1}{2} \sum_{i=1}^{b} k_i (l_i(\{x\}) \square L_i)^2$$

$$(1)$$

Using a Newton approach, minimization leads to an iterative expression (2) (see [2])

$$\left[\frac{\square F(\{x\})}{\square(\{x\})} \right]_{x_0} (\{x\} \square \{x_0\}) = [H(\{x\})](\{x\} \square \{x_0\})$$

$$(2)$$

In order to introduce P-joint restrictions, a kinematic restriction was proposed, leading to some problems to achieve convergence. To solve those problems, another approach can be used, based on assigning linear elastic behavior also to the P-joint. Thus, an extended expression of the energy of deformation can be derived (3).

$$\square(\{x\}) = \frac{1}{2}\sum_{i=1}^{b} k_i (l_i(\{x\}) \square L_i)^2 + \frac{1}{2}\sum_{j=1}^{d} q_j(d_j)^2$$

(3)

This minimization leads also to an expression similar to (2). This approach solves the convergence problems shown by using kinematic restrictions.

This error function has been successfully used for the optimum synthesis of mechanisms with a second order Newton minimization, only if a good starting mechanism is provided. For other mechanisms, which are far from a good solution, this type of model does not accurately represent their quality since in some cases, even for those mechanisms characterised by a low quality, the error may be too low. All of this invalidates the use of this function in genetic algorithms.

The here proposed error function calculates the quadratic addition of the minimum linear or angular distance to the synthesis positions which can be achieved by the mechanism.

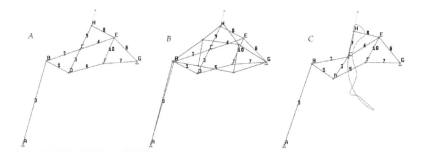

Figure 2. Minimum distance problem. A: Initial position. B: Solution. C: Real path described

In Figure 2 this error function is described for a synthesis point. The objective is the path described by node H verifies the point described with a cross. The minimum distance point is shown in Figure 2B. The path described by the mechanism is in Figure 2C.

In order to calculate these minimum distance positions, two different approaches can be used. The easiest one is to create a new mechanism by combination of the original mechanism and some fictitious elements related to the imposed constraint. These elements are given a stiffness comparatively small related to that of the original elements, thus being deformation of original elements much smaller than that of the fictitious ones. In the example shown in Figure 2 the fictitious bar would be a cero length bar that links node H with its target point, marked with a cross.

A more rigorous approach can be the use of Lagrange multipliers, minimizing the distance and using the fixed length of the trusses of the mechanism as restrictions (see [2]).

The second approach, even though it is more rigorous, has some problems of convergence when the initial position is far from the final solution. As a solid function is needed for genetic algorithms, the first simplification must be used, or at least a combination of both methods, in order to give Lagrange multipliers a good starting point.

Currently the first approach is used to test validity of genetic algorithms in solving this problem. The second approach will be used lately for improving the performance of the algorithm only if the first approach has been successful.

3.   OBTAINED RESULTS

In order to check the efficiency of the method, a set of experiments has been developed for a particular problem. This problem is the generation of a L-shaped path with a simple mechanism (a four bar mechanism).

Experimentation has been developed with a sensitivity analysis of 3 genetic algorithm parameters, with a total of 8 variable combination. Each combination has been repeated 10 times in order to check consistency. All parameters have been selected in the neighborhood of those obtained with a previous experimentation done with a similar error function. The parameters have been checked at 2 levels, being the following:
- Population size: 100/1000 individuals
- Mutation tax: 0,01/0,001
- Crossover tax: 0,5/0,65
  Non studied selected parameters:
- Selection operator: Tourney
- Crossover type: two point
- Population replacement: generational
- Elitism: none
- Convergence criteria: no best value change in 10 generations
- Genome codification: BCD (Gray)

The problem to optimise is the synthesis of a four bar mechanism verifying the path described by the points indicated in Table 1.

Table 1. Path points coordinates

25.0	15.0
27.5	15.0
30.0	15.0
32.5	15.0
35.0	15.0
35.0	17.5
35.0	20.0
35.0	22.5
35.0	25.0
35.0	27.5

As ten points are given, there is no exact solution for this problem, so the aim is to obtain a path as close as possible to that defined by those points. In order to define the search space, node codification has been limited to points in interval X [-200,200] Y [-200.200]. The resolution chosen is 20 bits per value. Results of the tests are shown in Table 1.

Table 2. Results of the tests

Population	100				1000			
Mutation Tax	0.01		0.001		0.01		0.001	
Crossover Tax	0.5	0.65	0.5	0.65	0.5	0.65	0.5	0.65
Best individual (average of 10 executions)	11.32	12.11	18.17	19.31	5.35	5.24	8.53	7.18
Best individual (absolute)	4.26	4.58	13.06	6.89	3.32	3.02	3.02	4.34
Iteration Number (average of 10 executions)	290.9	737.8	301.5	268.3	200.8	855.7	225	194.9
Number of executions without convergence	0	4	1	0	0	7	0	0

The best average results are between 20 and 5, it seems to be a small quantity, but actually there is a noticeable difference of quality in those results. Since the error function measures the quadratic addition of minimum distances from the real achievable path of the mechanism to the synthesis points, a value of around 20 indicates that the mechanism has only some points of its path close to some of the points of synthesis. As an example, in Figure 3,4,5,6 some results are shown.

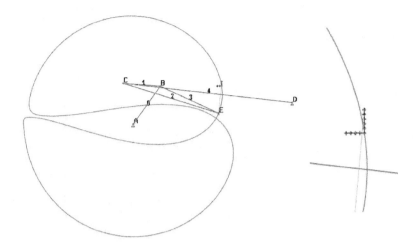

Figure 3. Obtained path for a mechanism yielding a fitness of 31.45 and detail

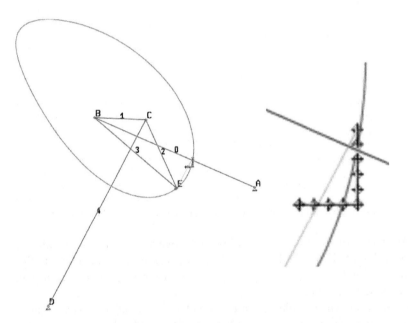

Figure 4. Obtained path for a mechanism yielding a fitness of 20.02 and detail

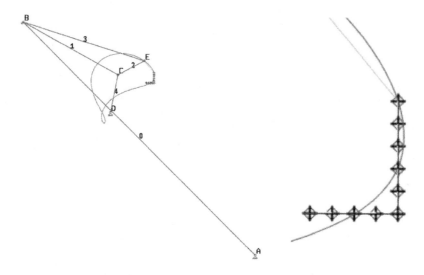

Figure 5. Obtained path for a mechanism yielding a fitness of 10.68 and detail

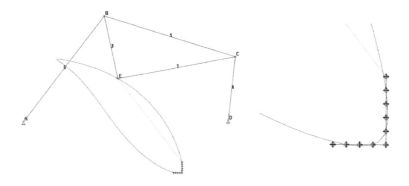

Figure 6. Obtained path for a mechanism yielding a fitness of 3.23 and detail

As it is shown, the quality difference between mechanisms yielding a fitness value of about 5 and of about 30 is noticeable. It can also be said that the function delivers results with a good relationship with the real quality of the generated mechanism.

As it is shown in table 1, with low values of population size, genetic algorithm does not give a good result, being this trend independent upon the mutation tax or the crossover rate.

An increase of mutation tax also leads to a increase in the quality of average solution. This can imply that exploration is of high importance in this problem, but there is also an increase in the number of generations with this increment in the mutation rate. This side effect can also contribute to the increase of the quality. In this case, convergence criterium should be changed.

Looking at the historical, evolution graphs built for the best individual history show too big oscillations with mutations of 0.01 and a very soft decrease of error with mutations of 0.001, the best results being probably between these values.

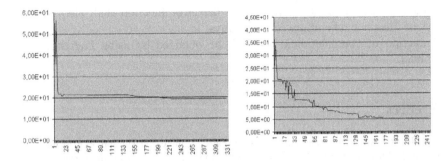

Figure 7. Evolution history for the best individual for evolutions with mutation rates of 0.001 and 0.1

In order to check the efficiency of the genetic algorithm, it has also been compared with a random search. This random search has been made with a search size of 1000000 individuals, this search has always a bigger cost than the most expensive of the tested genetic algorithm executions, and yielded a best individual with a fitness of 11.82, far from the worst individual obtained with the genetic algorithm with the combination of parameters that yielded the best results.

4.   CONCLUDING REMARKS AND FUTURE WORK

A new approach for the dimensional synthesis of mechanisms has been developed based on genetic algorithms. This approach has shown good behavior in solving complex synthesis problems, while introducing some advantages such as easy limitation of the design space or multiplicity of solutions.

In order to check the results obtained, a program (Mecano) has been developed, this program solves both initial and deformed positions and also generates paths for any mechanism with both R and P joints.

A genetic algorithms program has also been developed, based on David Levine's PGAPack and the error function proposed here. This implementation also includes a modification for its use with variable complexity function techniques, which reduce computational cost of the problem.

The error function has shown good behavior when using gray code for its codification. Absolute coordinates for the nodes are coded, so that both dimensions and configuration are fully defined. Currently its behavior with different combinations of parameters is studied, and also different genoma codifications are to be tested.

Some problems currently penalized in the error function shall also be approached in some other more efficient ways.

Reduction of computational cost will be obtained by using Lagrange multipliers.

References

[1] R. Avilés; G. Ajuria; E. Amezua; V. Gomez-Garay, A finite element approach to the position problems in open-loop variable geometry trusses, *Finite elements Anal. Design,* Vol.34, 2000, pp.233-255

[2] R. Avilés, G. Ajuria, A. Bilbao, J Vallejo, Lagrange multipliers and the primal-dual meted in the nonlinear static equilibrium of multibody systems, Commun. Numer. Math. Eng., Vol.14, 1998, pp.463-472

[3] J.M. Jiménez; G. Álvarez; J. Cardenal; J. Cuadrado, A Simple and General Method for Kinematic synthesis of spatial mechanisms, *Mechanism and Machine Theory,* Vol.32, No.3, 1997, pp. 323-341

[4] J.R. McGarva, Rapid search and selection of path generating mechanisms from a library, *Mech. Mach. Theory,* Vol 29, No.2, 1994, pp.223-235.

[5] C. Zhang, R.L. Norton, T. Hammond, Optimization of parameters for specified path generation using an atlas of coupler curves of geared five bar linkages, *Mech. Mach. Theory* Vol 19, 1984, pp. 459-466.

[6] A. Vasiliu; B. Yannou, Dimensional synthesis of planar mechanisms using neural networks, application to path generator linkages, *Mechanism and Machine Theory*, Vol 36, No 2, 2001, pp. 299-310

EVOLUTIONARY COMPUTATION IN THERMOELASTIC PROBLEMS

Adam Długosz

Department for Strength of Materials and Computational Mechanics
Silesian University of Technology, Gliwice, Poland

adlugosz@rmt4.kmt.polsl.gliwice.pl

Abstract: The paper deals with the application of evolutionary algorithms (EA) and the boundary element method (BEM) in optimization and identification of elastic structures under thermomechanical loading. A shape optimization problem was solved for various thermomechanical criteria with upper bound on the volume of the structure. The identification of voids basing on the information about measured displacements and temperatures in boundary sensor points was also considered. This problem was solved as well as for both ideal deterministic and randomly disturbed values of measured displacements and temperatures. Several tests and practical examples of optimization and identification were presented.

Keywords: thermoelastic problem, boundary element method, evolutionary algorithms, shape optimization, evolutionary optimization

1.   INTRODUCTION

The solution of shape optimization problems as well as identification problems of structures being under the thermo - mechanical loading need a computational technique, which enables to solve the direct boundary - value problem of thermoelasticity in a very effective manner. The boundary element method occurs to be the numerical method which is very convenient for such problems. Coupling of the boundary element method with sensitivity analysis methods has enabled the solutions of various shape optimization problems of thermoelasticity. The main drawback of such approach is the fact that only a local optimal solution can be obtained. An alternative to an such approach is the coupling of the boundary element method with evolutionary algorithms. Applications of evolutionary algorithms in optimization problems give a great probability of finding the global optimal solution. The paper

69

T. Burczyński and A. Osyczka (eds),
IUTAM Symposium on Evolutionary Methods in Mechanics, 69–80.
© *2004 Kluwer Academic Publishers.*

deals with the application of evolutionary algorithms (EA)[9] and the boundary element method (BEM)[1] [2] in the optimization and identification of elastic structures under thermomechanical loading. This work is an extension of previous papers in which the coupling of EA and BEM has been used in the generalized shape optimization of elastic structures [5], the optimization of cracked structures [3], the shape optimization of elastoplastic structures [6] [7] and the identification of voids and cracks [4]. In order to solve the optimization and inverse problems the evolutionary algorithms are proposed. To minimize the number of design parameters the shape of the boundary is modelled by Bezier curves.

2.   EVOLUTIONARY OPTIMIZATION OF THERMOELASTIC STRUCTURES

The evolutionary algorithm can be considered as the modified and generalized genetic algorithm in which population are coded by floating point reprezentation[9]. Fig.1 shows the main steps of the evolutionary algorithm. The solution of this problem is given by the best chromosome whose genes represent design parameters responsible for the shape of the structure or parameters of the internal voids. The evolutionary algorithm starts with a population of chromosomes randomly generated from the feasible solution domain.

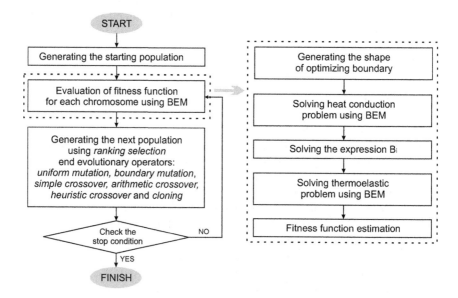

Figure 1.   The evolutionary algorithm of the thermoelastic structures optimization.

Two kinds of mutation operators are applied: an uniform mutation and a boundary mutation. The operator of the uniform mutation replaces a gene of the chromosome with a new random value. The operator of the boundary mutation is a special type of the uniform mutation. The gene after mutation receives one of the boundary value (left or right) with the same probability. The boundary mutation is very useful for problems in which the result of the solution is on the boundary or very near of the boundary.

Three kinds of crossover operators are applied: a simple crossover, an arithmetical crossover and a heuristic crossover. Generally the crossovers combine the genetic material inside the population. The simple crossover creates a pair of offspring chromosomes depending on two parent chromosomes. The offspring contains some genes from one and some from other parent. The operator of the arithmetic crossover creates two new chromosomes as a linear combination of parents chromosomes. The heuristic crossover uses values of the fitness function to create new chromosomes.

The operator of the cloning increases the probability of surviving the best chromosome by duplicating this one to the next generation. The ranking selection allows chromosomes with the great value of the fitness function to survive. The first step of the ranking selection is sorting all the chromosomes according to the value of the fitness function. Then on the basis of the position in the population the probability of surviving is attributed to the every chromosome.

3.   EVALUATION OF THE FITNESS FUNCTION BY MEANS OF THE BOUNDARY ELEMENT METHOD

3.1   Boundary element method in thermoelasticity problem

In order to evaluate the fitness function one needs to solve the thermoelasticity problem. One of the numerical method which enables finding the solutions is the boundary element method. It is assumed that uncoupled steady state thermoelasticity in which the strain field depends on the temperature field but temperature field does not depend on the strain field.

Boundary integral equations for the BEM in thermoelasticity can be express as follows:

$$c_{ij}u_j + \int_\Gamma p_{ij}^* u_j d\Gamma = \int_\Gamma u_{ij}^* p_j d\Gamma + \int_\Omega u_{ij,j}^* \gamma T d\Omega \qquad (1)$$

where u is a field of boundary displacements, p is a field of tractions, u_{ij}^* and p_{ij}^* are fundamental solution of elastostatic problem.

The domain integral which depends on temperature can be transformed into a boundary integral using the stress function introduced by Papkovich and Neuber. Then the domain integral has the form:

$$B_i = \int_\Gamma P_i T d\Gamma - \int_\Gamma Q_i q d\Gamma \tag{2}$$

where P_i and Q_i are are functions which depend on the dimension of the problem.

In order to find the boundary temperature field T and its normal derivative q in expression (2) it is necessary to solve the boundary integral equation for the heat conduction problem given by:

$$cT = \int_\Gamma T^* q d\Gamma - \int_\Gamma Q^* T d\Gamma \tag{3}$$

where T^* and Q^* are fundamental solution of the heat conduction problem.

The boundary integral equations (1) and (3) after discretization are transformed into algebraic equations:

$$\begin{aligned} ST &= Rq \\ Hu &= Gp + B \end{aligned} \tag{4}$$

where column matrices T, q, u and p contain boundary values of temperatures, heat fluxes, displacements and tractions.

The column matrix B depends on the temperature and the heat flux on the boundary. Elements of matrices S, R, H and G can be obtained by evaluating boundary integrals. All unknowns in matrices can be passed to left-hand side and finally one obtains:

$$\begin{aligned} KY &= Q \\ AX &= F \end{aligned} \tag{5}$$

where X is a vector of unknowns $u's$ and $p's$ boundary values, Y is a vector of unknowns $T's$ and $q's$ boundary values. F and Q are found by multiplying the corresponding columns by known values of $u's$, $p's$ and $T's$, $q's$.

3.2   Optimization criteria

In the case of the shape optimization problem two criteria are proposed:

- minimum global compliance of thermoelastic structures

$$F = \int\limits_{\Gamma} \left(\frac{1}{2} p \cdot u\right) d\Gamma \qquad (6)$$

where u is a field of boundary displacements, p is a field of tractions

- minimum equivalent stresses on the boundary

$$F = \int\limits_{\Gamma} \left(\frac{\sigma_{eq}}{\sigma_0}\right)^n d\Gamma \qquad (7)$$

where σ_{eq} are equivalent stresses on the bounary, σ_0 is a reference stress and n is a natural number.

For the identification of an unknown inner boundaries the sensor points on the external boundary were used. This problem is solved by minimization of the following functional:

$$F = \delta \sum_{k=1}^{M} (u^k - \hat{u}^k)^2 + \eta \sum_{l=1}^{N} (T^l - \hat{T}^l)^2 \qquad (8)$$

where $\hat{u}^k$ and $\hat{T}^l$ are measured values of displacements and temperatures, u^k and T^l are computed values of displacements and temperatures in boundary points k and l respectively, δ and η are weight coefficients, M, N are numbers of sensors.

Measured values are simulated numerically by solving the boundary value problem of thermoelasticity by the BEM for the actual position of internal voids. This problem has been solved for both ideal deterministic (no noise) and randomly disturbed (with noise) values of measured displacements and temperatures. For non ideal deterministic measured values the Gaussian distribution was applied. The density function $N(\mu, \sigma)$ is shown in Fig.2, where $q = \hat{u}^k$ or $q = \hat{T}^l$.

The expected value μ is equal to the ideal deterministic value of measured displacements or measured temperatures. The standard deviations is equal to $\frac{1}{3}$ of the maximal error. The maximal error of measurement is 10 percent [8].

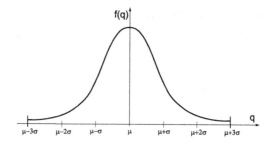

Figure 2.   The density function of the Gaussian distribution $N(\mu, \sigma)$.

4.   NUMERICAL EXAMPLES

2D structures modelled in plain strain state is considered. Table 1 contains parameters of evolutionary algorithm and Table 2 contains material parameters applied for every numerical test.

Table 1.   Parameters of evolutionary algorithm

Probability of uniform mutation	0.03
Probability of boundary mutation	0.02
Probability of simple crossover	0.10
Probability of arithmetical crossover	0.10
Probability of heuristic crossover	0.10
Probability of cloning	0.03
Selection coefficient	0.2

Table 2.   Material parameters

Shear modulus	80 GPa
Poisson ratio	0.23
Coefficient of thermal exp.	$12,5 \cdot 10^{-6} \frac{1}{K}$

Example 1

A shape optimization of the structure presented in Fig.3(a) is considered. Only the parts of the boundary, where temperature field T_2^0 and heat flux q^0 are prescribed, undergo the shape modification. The Bezier curves consist of 5 control points are used to model the optimized boundary. The method of the modeling of the optimized boundaries is presented in Fig.3(b). The possible position of the control points of the Bezier curves are inside the dotted rectangles. For the sake of the symmetry total number of design parameters is equal to 11.

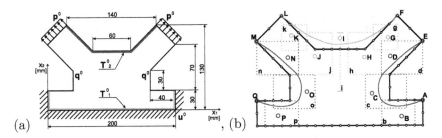

Figure 3. (a) A structure under thermomechanical loading, (b) Modelling of the optimized boundaries

The objective of shape optimization is the minimization of boundary equivalent stresses (7) with imposed constrains on the upper bound of the volume of the structure. Table 3 contains values of the boundary conditions. The optimal shape of the structure for different values of the temperature T_1^0 is presented in Fig.4

Table 3. Values of the boundary conditions

Number of chromosomes	200
Number of iterations	500
Number of design parameters	11
u^0	0
q^0	0
T_1^0	$0°C$; $50°C$; $100°C$; $200°C$
T_2^0	$0°C$
p^0	$500kN/m$

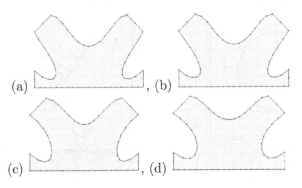

Figure 4. Results for $T_1^0 =$: (a) $0°C$, (b) $50°C$, (c) $100°C$, (d) $200°C$,

Example 2

A square plate with circular voids is considered (Fig.5). For the sake of the symmetry only a quarter of the structure is taken into consideration. Table 4 contains values of the boundary conditions. This example was solved for the following two cases:

- The optimal positions and radii of two internal voids were searched

- The positions, radii and number of the voids (form 1 to 5) were searched.

The objective of the shape optimization is the minimization of the radius displacement (6) on the boundary where traction p^0 are prescribed.

Table 4. Values of the boundary conditions

Number of chromosomes	100
Number of iterations	200
Number of design parameters	3 or 5
u^0	0
q^0	0
T_1^0	$500°C$
α_1	$1000W/m^2K$
T_2^0	$10°C$
α_2	$1000W/m^2K$
T_1^0	$50°C$
α_0	$20W/m^2K$

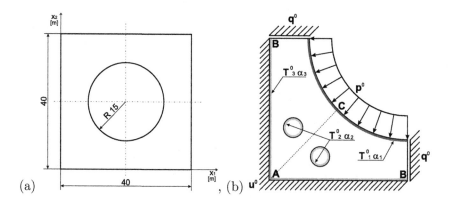

(a) , (b)

Figure 5. (a) A square plate with circular void, (b) Boundary conditions for the structure

The optimal position and size of the voids for the considered cases is presented in Fig.6. Tables 5a and 5b contains parameters of the internal boundaries and values of the fitness function after optimization.

Table 5a. Parameters of the voids and value of the fitness function for constant number of the voids

x_1 coordinate	12, 02
x_2 coordinate	3, 47
radius	1, 89
value of the fitness function	1842, 25

Table 5b. Parameters of the voids and value of the fitness function for changeable number of the voids

x_1 coordinate of the void 1	13, 80
x_2 coordinate of the void 1	4, 15
radius of the void 1	0, 97
$x_1 = x_2$ coordinate of void 2	8, 08
radius of the void 2	0, 59
value of the fitness function	1102, 58

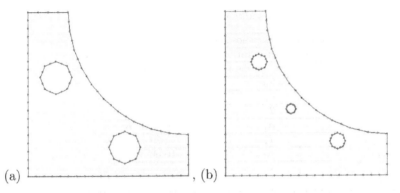

(a)   , (b)

Figure 6. The results of the optimization for: (a) 2 internal voids, (b) changeable number of voids

Example 3

The structure with three internal voids, shown in the Fig.7(a), was considered. The boundary conditions values and positions of the sensor points are shown in Fig.7(b). The positions, radii and number of the voids (form 1 to 5) were searched. Table 6 contains the values of the boundary conditions.

The numerical tests were performed for 3 cases with noise and for 3 cases without noise:

- 30 temperature sensor points and 28 displacement sensor points,

- 58 temperature sensor - located in every sensor points,

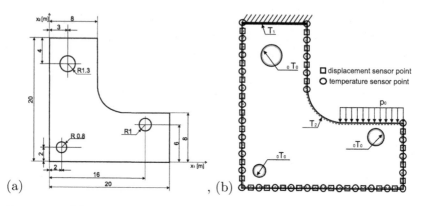

(a) , (b)

Figure 7. (a) A structure with circular voids, (b) Boundary conditions for the structure

Table 6. Values of the boundary conditions

Number of chromosomes	500
Number of iterations	300
Number of design parameters	15
T_0	$100^\circ C$
α_0	$1000W/m^2K$
T_1	$0^\circ C$
α_1	$20W/m^2K$
T_2	$20^\circ C$
p_0	$100MN/m$

- 58 displacement sensor - located in every sensor points.

Example results of identification without noise are shown in Fig.8, and Fig.9 contains numerical results of identification with noise for 3 above cases.

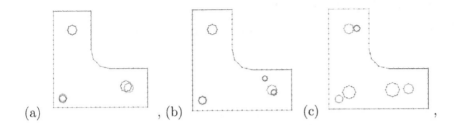

Figure 8.   Results of identification without noise: (a) 30 temperature sensor points and 28 displacement sensor points, (b) 58 temperature sensor points, (c) 58 displacement sensor points

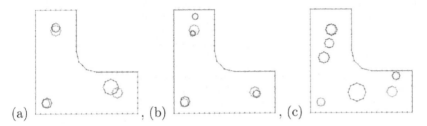

Figure 9.   Results of identification with noise: (a) 30 temperature sensor points and 28 displacement sensor points, (b) 58 temperature sensor points, (c) 58 displacement sensor points

5.   CONCLUDING REMARKS

Coupling of the BEM and evolutionary computation is a very efficient approach to solving shape optimization and inverse thermoelastic problems. Using Bezier curves reduces the number of design parameters and decreases the time of computation. The application of evolutionary algorithms increases the probability of finding the global solutions. In the case of use gradient methods based on sensitivity analysis finding the global solution depends on the starting point. The evolutionary algorithm performs multidirectional optimum searching by exchanging information between chromosomes and allows to survive best chromosomes. One drawback of evolutionary methods is the time-consuming calculation.

Preformed tests of identification show that having measured coupled fields of displacements and temperatures the evolutionary identification of voids is more effective than having only displacements or temperatures separately.

ACKNOWLEDGEMENT

The research was carried out in the framework of the KBN grant 4T11F00822.

References

[1] Banerjee P.K., The Boundary Element Methods in Engineering, *McGRAW-HILL Book Company Europe*, Berkshire 1994.

[2] Brebbia C.A., Dominiguez J., Boundary Elements An IntroductoryCourse, *Computational Mechanics Publications*, Southampton Boston 1989.

[3] Burczyński T., Beluch W., Kokot G., Optimization of cracked structures using boundary elements and evolutionary computation, *In: Boundary Element Techniques*, (ed. M.H.Aliabadi), London, 1999.

[4] Burczyński T., Beluch W., Długosz A., Kokot G., Kuś W., Orantek P., Evolutionary BEM computation in shape optimization problems, IU-TAM/IACM/IABEM Symposium on Advanced Mathematical and Computational Mechanics Aspects of the Boundary Element Method, ed. Burczynski T., Kluwer, 2001. London, 1999.

[5] Burczyński T., Kokot G., The evolutionary optimization using genetic algorithms and boundary elements, Proc. 3rd World Congress on Structural and Multidisciplinary Opimization, Buffalo, New York, USA, 1999.

[6] Burczyński T., Kuś W., Evolutionary methods in shape optimization of elasto-plastic structures, 33rd Solid Mechanics, Zakopane 2000.

[7] Burczyński T., Kuś W., Distributed evolutionary algorithms in shape optimization of nonlinear structures, 4th International Conference, PPAM, Naleczow 2001.

[8] Długosz A., Boundary element method in sensitivity analysis and optimisation of thermoelastic structures, PhD thesis, Silesian University of Technology, Gliwice 2001.

[9] Michalewicz Z., Genetic Algorithms + Data Structures = Evolutionary Programs, *Springer-Verlag*, AI Series, New York, 1992.

MANAGEMENT OF EVOLUTIONARY MAS FOR MULTIOBJECTIVE OPTIMISATION

Grzegorz Dobrowolski
Marek Kisiel-Dorohinicki
Department of Computer Science
AGH University of Science and Technology, Kraków, Poland
{grzela,doroh}@agh.edu.pl

Abstract: In the paper an agent-based evolutionary approach to searching for a global solution in the Pareto sense to multiobjective optimisation is discussed. The main stress is put on problems of effective management of such a system. Management mechanisms based on closed circulation of life energy that sustain autonomy of agents and allow for control of the dynamics of agent population are proposed. Preliminary experimental results conclude the work.

Keywords: evolutionary multi-objective optimisation, management of multi-agent systems

1. INTRODUCTION

A solution to a multiobjective optimisation problem (in the Pareto sense) means the determination of all nondominated alternatives from the set of possible (feasible) decision alternatives (options) defined by a system of inequalities (constraints) and equalities (bounds). Such alternatives create a set of continuum power called sometimes the Pareto frontier. In a general case effective approximation of the Pareto set is hard to obtain. For specific types of criteria and constraints (e.g. of linear type) some methods are known, but even in low-dimensional cases they need much computational effort. Fortunately for many complex problems an evolutionary approach may be successfully applied [11].

For the last 20 years a variety of evolutionary multicriteria optimisation techniques have been proposed [3, 4]. Early approaches have not used directly the information about domination relation between solutions. Conversely, in most recently presented methods some sort of ranking is used, reflecting the "degree" of domination of particular solutions.

T. Burczyński and A. Osyczka (eds),
IUTAM Symposium on Evolutionary Methods in Mechanics, 81–90.
© *2004 Kluwer Academic Publishers.*

This allows for better approximation of the Pareto set, especially when supported with some niching techniques (like fitness sharing), which prevent genetic drift and enable sampling of the whole frontier.

The proposed agent-based approach indirectly utilizes the information about domination via energetic punishment/reward mechanism, which allows for intensive exploration of the search space, and effective approximation of the whole Pareto set [7, 9]. Yet in such systems the problems of agent populations management become of vast importance. A typical parameter which must be considered in this context is the total number agents in population(s). Particularly, increasing the number of agents usually improves the ability of the system to perform its tasks, but at the same time may cause the whole system to work inefficiently (e.g. because of excessive load of executed actions) or even prove impractical in real applications. The problem is that because of assumed autonomy of agents direct control of the system seems impossible. In this situation only indirect mechanisms, based on analogies to (simulating) the ones from the real world (nature) may be applied [10].

The paper starts with a short characteristics of multi-agent systems dedicated for computational applications (section 2). Next (section 3) comes a short presentation of the idea of agent-based evolutionary computation and the biological-like mechanisms of management of evolving agent populations. Section 4 contains a description of **EMAS** for multi-objective optimisation. Experimental studies illustrating the proposed mechanisms of management are reported in section 5.

2.   MASS MULTI-AGENT SYSTEMS

Recently an increasing attention is paid to hybrid soft computing systems, i.e. systems employing a variety of methodologies (e.g. neuro-fuzzy or fuzzy-genetic), which is possible because particular methods are complementary rather then competitive [1]. In this context the idea of an autonomous agent and agent-based system seems interesting and promising [8].

As it is commonly regarded, multi-agent systems are ideally suited to representing problems that have multiple problem solving methods, multiple perspectives or multiple problem solving entities [5, 12]. Rough analysis of the characteristics of **MAS** drives to the conclusion that it can be considered also as a means for modelling social behaviour or mechanisms. Agents are forced to co-operate. They have to communicate in order to establish a common goal, plan their common activities and coordinate how a plan is realised. Moreover, there can exist no strict agreement with respect to it. Then, it is better to say that agents have their private aims and organisation of the system (population of them) makes searching for private satisfaction a way the common goal to be fulfilled.

Yet another term has to be recalled here – agent-based system. In principle, an agent-based system is conceptualised in terms of agents, but implemented even without any software structures corresponding to them. Such abstract MAS is suitable in applications that locate in the area of soft computing [10].

A multi-agent system consists of the set of *agents* (ag $\in$ Ag) and some *environment* (env) they live in:

$$\mathsf{mas} \equiv \langle \mathsf{Ag}, \mathsf{env} \rangle \tag{1}$$

The environment may have spatial structure and contain some *information* and/or *resources*, which may be observed by agents.

The functionality of i-th agent is defined by a set of *actions* (act $\in$ Act$_i$) it is able to perform, depending on its *state* (stat$_i$):

$$\mathsf{ag}_i \equiv \langle \mathsf{Act}_i, \mathsf{stat}_i \rangle \tag{2}$$

The action is an atomic (indivisible) activity, which may be executed by an agent in the system.

Specifically, a **mass multi-agent system** consists of a relatively large number of often not so complicated agents [10]. Operation of mMAS is aimed at a well-defined goal, which emerges together with organisation among the agents as a result of performed actions. Because of both computational simplicity and a huge number of the agents, influence of the single agent's behaviour on the overall system operation may be neglected. Although it is not required in general, agents of a mass system are often identical in the sense of an algorithm and built-in actions. To fulfil the characteristics of mass multi-agent systems, it is necessary to present basic mechanisms that can be incorporated in a system of that kind:

Generation and removal of agents corresponds to a type of *open multi-agent systems*, in which a number of agents changes during operation.

Information exchange among agents or between agents and the environment (if such is implemented).

Exchange of resources other than information is also possible. Resources may be associated with each action performed (necessary energy expenditure), or there may be actions intended to resource manipulation.

Migration of agents needs a structure embedded in the environment that reflects spatial phenomena of mMAS (e.g. a graph).

Implementation of a mechanism means to equip agents with an appropriate set of actions and, if it is necessary, extend the environment with a corresponding feature or structure. In general realisation of each action changes not only the internal state of the acting agent but also states of the other agents and the state of the environment:

$$\text{act}: \quad \text{mas} \rightarrow \text{mas}^* \tag{3}$$

where mas^* denotes a new state of the system (i.e. after the action act is performed).

3.   EVOLUTIONARY mMAS AND THE IDEA OF MANAGEMENT VIA LIFE ENERGY

An *evolutionary* multi-agent system (EMAS) is a kind od mMAS, in which basic mechanisms depicted above are designed so that evolutionary phenomena emerge at a population level [2, 8, 7]. This means that agents are able to *reproduce* (generate new agents) and may *die* (be eliminated from the system) realising the phenomena of *inheritance* and *selection*. Inheritance is to be accomplished by an appropriate definition of reproduction, which is similar to classical evolutionary algorithms. A set of parameters describing basic behaviour of an agent is encoded in its genotype, and is inherited from its parent(s) – with the use of mutation and recombination.

The proposed principle of selection corresponds to its natural prototype and is based on the existence of non-renewable resource called *life energy*, which is gained and lost when agents execute actions. Increase in energy may be considered a reward for 'good' behaviour of an agent, decrease – a penalty for 'bad' behaviour (of course which behaviour is considered 'good' or 'bad' depends on the particular problem to be solved). At the same time the level of energy determines actions an agent is able to execute. In particular, low energy level should increase possibility of death and high energy level should increase possibility of reproduction.

Considering its computational properties EMAS may be regarded as a new search and optimisation technique utilising a *decentralised* model of evolution (its more detailed description may be found e.g. in [6]).

The key idea of the described mechanisms of management is to make EMAS flexible with respect to the number of agents so that the population would be limited but could automatically grow when it is necessary.

Automatically means releasing a user from direct management of the course of computation [10].

The mechanism is built as follows: *life energy* eng is distinguished as one of the components of stat — the state of an agent:

$$\text{stat} = \langle \vec{x}, \text{eng} \rangle \tag{4}$$

The rest of components represent the problem to be solved by the system as well as the behaviour of an agent. At the beginning of EMAS run each agent is given the same amount of energy $\text{eng}_{\text{ag}}(0) = \text{eng}_0 > 0$. Next, energy is gained or lost when agents execute actions. Core actions of generation and removal of agents mechanism can be described consecutively:

- action of *death* (die):

$$\text{die}: \quad \langle \text{Ag}, \text{env} \rangle \rightarrow \langle \text{Ag} \setminus \{\text{ag}\}, \text{env} \rangle \tag{5}$$

- action of *reproduction* (rp):

$$\text{rp}: \quad \langle \text{Ag}, \text{env} \rangle \rightarrow \langle \text{Ag} \cup \{\text{ag}^*\}, \text{env} \rangle \tag{6}$$

where the energy of the new (offspring) agent ag^* is:

$$\text{stat}^* = \langle \vec{x}^*, \text{eng}_0 \rangle \tag{7}$$

where $\vec{x}^*$ denotes partial representation of the problem to be solved inherited with mutation from $\vec{x}$, and eng_0 — initial energetic level.

For a parental agent the following holds:

$$\text{ag} = \langle \vec{x}, \text{eng} - \text{eng}_{\text{rp}} \rangle \tag{8}$$

The energy balance for an agent along a given period of time is as follows:

$$\text{eng}(t_{k+1}) = \text{eng}(t_k) + \Delta\text{eng}^+(t_k) - \Delta\text{eng}^-(t_k) \tag{9}$$

where: t_{k+1}, t_k – moments, when the state of the system is observed,

$\Delta\text{eng}^+(t_k), \Delta\text{eng}^-(t_k)$ – total energy gained and lost, respectively, in time period (t_k, t_{k+1}).

The level of life energy possessed simultaneously determines actions an agent is able to execute. Thus the strategy of selection should be defined so as high energy level should increase the probability of reproduction — generation of offspring agents: $p_{\text{rp}} = p_{\text{rp}}(\text{eng})$, and low energy level should increase the probability of death — elimination of the agent: $p_{\mathsf{die}} = p_{\mathsf{die}}(\mathsf{eng})$.

The total amount of life energy can be used as a means for management of this kind of EMAS in two possible regimes.

- Operation of the system may be modified by adding or taking away a portion of the energy. *Power supply* changes.

- Closed circulation of life energy in the system is realised, which makes that the total energy possessed by all agents and in the environment remains constant during the whole run.

Of course, assurance of the closed circulation of energy needs each change in the agent's energy to be reflected by an appropriate change in energy of another agent or the environment.

4.   EMAS FOR MULTIOBJECTIVE OPTIMISATION

In order to find the approximation of Pareto frontier for a given multicriteria optimisation problem, agents of EMAS must act according to the energetic reward/punishment mechanism, which prefers nondominated agents [7, 9]. This is done via so-called *domination principle*, forcing dominated agents to give a fixed amount of their energy to the encountered dominants. This may happen, when two agents communicate with each other and obtain information about their quality with respect to each objective function. Thus this mechanism consists of several actions of communication and the action of energy transfer and may be compactly described as:

$$\mathsf{tr} : \quad \mathsf{mas}|_{\mathsf{ag}_1,\mathsf{ag}_2} \to \mathsf{mas}|_{\mathsf{ag}_1^*,\mathsf{ag}_2^*} \tag{10}$$

where states of agents ag_1 and ag_2 change according to *domination principle*:

$$\mathsf{stat}_1 = \langle \vec{x}_1, \mathsf{eng}_1 \rangle \qquad \mathsf{stat}_1^* = \langle \vec{x}_1, \mathsf{eng}_1 - \Delta\mathsf{eng} \rangle$$
$$\mathsf{stat}_2 = \langle \vec{x}_2, \mathsf{eng}_2 \rangle \qquad \mathsf{stat}_2^* = \langle \vec{x}_2, \mathsf{eng}_2 + \Delta\mathsf{eng} \rangle$$

and $\vec{x}_2$ dominates $\vec{x}_1$.

The flow of energy connected with the domination principle causes that dominating agents are more likely to reproduce whereas dominated ones are more likely to die. This way, in successive generations, nondominated agents should make up better approximations of the Pareto frontier.

The above description is only a general scheme of how to realise EMAS for multicriteria optimisation, to make the method operational for a particular problem the following issues must be considered:

- *Types of problems covered.* Both representation of solutions and variation operators should reflect the problem to be solved.

- *Dimensionality.* A large number of decision variables and criteria functions does not pose a problem for evolutionary optimisation techniques, only a suitable size of population and agent space should be selected. What is more, the evaluation of agents (domination principle) has linear complexity with respect to the number of criteria in consideration.

- *Quality of approximation.* Precision of the approximation of the Pareto frontier depends obviously on the population size and may be influenced this way. A harder problem is to ensure the uniform distribution of nondominated agents (solutions) along the frontier.

- *Incoherent or discontinuous Pareto frontier.* For a problem with an incoherent (discontinuous) Pareto frontier controlling the distribution of nondominated agents occurs even crucial, e.g. for discovering the gaps of certain dimension. The mechanism of *crowd* [9] is the proposed solution to the problem.

- *Dynamic changing of the problem.* The proposed method naturally allows modification of the problem during solving process that can be valuable features for some kind of applications.

5. EXPERIMENTAL STUDIES

The implementation was based on the *AgWorld* platform – a software framework based on PVM facilitating agent-based implementations of distributed evolutionary computation systems [2].

Experimental studies have been carried out for a simple (yet interesting, since leading to a discontinuous Pareto frontier) multiobjective optimisation problem [9]:

$$f_1(x, y, z, t) = -(x - 2)^2 - (y + 3)^2 - (z - 5)^2 - (t - 4)^2 + 5$$
$$f_2(x, y, z, t) = \frac{\sin x + \sin y + \sin z + \sin t}{1 + (\frac{x}{10})^2 + (\frac{y}{10})^2 + (\frac{z}{10})^2 + (\frac{t}{10})^2}$$

Below some results characteristic for management by means of life energy (self-management) are presented in short.

Firstly, Fig. 1 shows how the quality of solutions (understood as average distance between the model solution and these obtained during

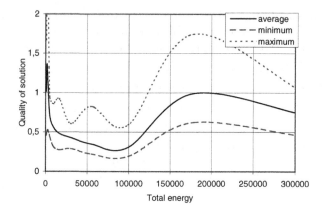

Figure 1.   Influence of total energy of the system on the quality of solutions obtained

studies) depends on the total energy of the system. It can be observed
that too little energy cannot guarantee a success, while the excess of it
can lead to the deterioration of the best obtained solution.

Secondly, the dependency of the number of computation steps on
the total energy (Fig. 2) indicates that too much energy unnecessary
lenghtens the process of finding a solution.

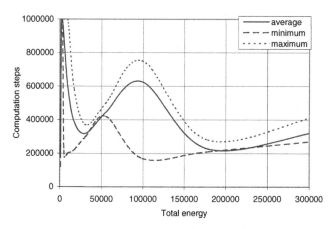

Figure 2.   Influence of total energy of the system on the total computation time

A similar situation is shown in Fig. 3 with respect to the number of
Pareto-optimal agents – reflecting accuracy of the Pareto frontier approximation. Deterioration of the result may be also observed for the

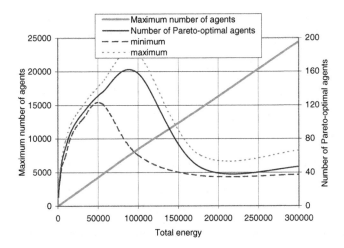

Figure 3.   Influence of total energy of the system on the maximum number of agents
and number of Pareto-optimal agents

badly chosen amount of life energy. At the same time, increasing the
total energy causes linear growth of the maximum number of agents during computation, which is due to the death threshold of energy defined
for an agent.

Additionally, it can be noticed that maxima and minima of the reproduced curves occur at rather close values of energy. This allows to
formulate the important practical indication that the amount of energy
as a parameter can and ought to be tuned each time a new optimisation
problem is solved.

In general, the presented observations confirm that the discussed management mechanism works and is effective. Of course, deeper analysis
needs further experiments.

6.   CONCLUDING REMARKS

Attaining utilitarian goal of mMAS and assuring effectiveness of that
process can be achieved merely by applying specific management mechanisms sustaining autonomy of individual agent actions. The results obtained for EMAS for multiobjective optimisation confirm legitimacy of
the general approach and effectiveness of the proposed scheme of management based on closed circulation of life energy.

Future research in the subject will concentrate on further theoretical
analysis of EMAS as well as utilization of different management mechanisms in various applications. Detailed studies will be carried out in

order to choose finally the built-in mechanisms, tune them, and compare the method with other known in the soft computing field.

ACKNOWLEDGMENTS

This work was partially sponsored by State Committee for Scientific Research (KBN) grant no. 4 T11C 027 22.

References

[1] P. Bonissone. Soft computing: the convergence of emerging reasoning technologies. *Soft Computing*, 1(1):6–18, 1997.

[2] A. Byrski, L. Siwik, and M. Kisiel-Dorohinicki. Designing population-structured evolutionary computation systems. In T. Burczyński, W. Cholewa, and W. Moczulski, editors, *Methods of Artificial Intelligence (AI-METH 2003)*, pages 91–96. Silesian University of Technology, Gliwice, Poland, 2003.

[3] C. A. Coello Coello, D. A. Van Veldhuizen, and G. B. Lamont. *Evolutionary Algorithms for Solving Multi-Objective Problems*. Kluwer Academic Publishers, 2002.

[4] K. Deb. *Multi-Objective Optimization using Evolutionary Algorithms*. John Wiley & Sons, 2001.

[5] J. Ferber. *Multi-Agent Systems. An Introduction to Distributed Artificial Intelligence*. Addison-Wesley, 1999.

[6] M. Kisiel-Dorohinicki. Agent-oriented model of simulated evolution. In W. I. Grosky and F. Plasil, editors, *SofSem 2002: Theory and Practice of Informatics*, Lecture Notes in Computer Science. Springer-Verlag, 2002.

[7] M. Kisiel-Dorohinicki, G. Dobrowolski, and E. Nawarecki. Evolutionary multi-agent system in multiobjective optimisation. In M. Hamza, editor, *Proc. of the IASTED Int. Symp.: Applied Informatics*. IASTED/ACTA Press, 2001.

[8] M. Kisiel-Dorohinicki, G. Dobrowolski, and E. Nawarecki. Agent populations as computational intelligence. In L. Rutkowski and J. Kacprzyk, editors, *Neural Networks and Soft Computing*, Advances in Soft Computing, pages 608–613. Physica-Verlag, 2003.

[9] M. Kisiel-Dorohinicki and K. Socha. Crowding factor in evolutionary multi-agent system for multiobjective optimization. In H. R. Arabnia, editor, *Proc. of Int. Conf. on Artificial Intelligence (IC-AI 2001)*. CSREA Press, 2001.

[10] E. Nawarecki, M. Kisiel-Dorohinicki, and G. Dobrowolski. Organisations in the particular class of multi-agent systems. In B. Dunin-Keplicz and E. Nawarecki, editors, *From Theory to Practice in Multi Agent Systems*, volume 2296 of *Lecture Notes in Artificial Intelligence*. Springer-Verlag, 2002.

[11] A. Osyczka. *Evolutionary Algorithms for Single and Multicriteria Design Optimization*. Physica Verlag, 2002.

[12] G. Weiss, editor. *Multiagent Systems: A Modern Approach to Distributed Artificial Intelligence*. The MIT Press, 1999.

PAMUC : A NEW METHOD TO HANDLE WITH CONSTRAINTS AND MULTIOBJECTIVITY IN EVOLUTIONARY ALGORITHMS

R. Filomeno Coelho[1], PH. Bouillard[1] and H. Bersini[2]
[1]*Continuum Mechanics Department & [2]Iridia, UniversitØLibre de Bruxelles, Belgium*

Abstract: Multiobjective optimization using evolutionary algorithms (EA) has become a wide area of research during these last years. However, only a few papers deal with the handling of preferences. To take them into account, a method based on multi-criteria decision aid, PROMETHEE II, has been implemented. Further-more, as the handling of the constraints is very critical in EA, an original approach (*PAMUC □Pr eferences Applied to MUltiobjectivity and Constraints*) has been proposed, which considers the satisfaction of the constraints as another objective.

Keywords: evolutionary algorithms, constraints, multiobjective, preferences, multicriteria.

1.   INTRODUCTION

The ability of evolutionary algorithms to explore widely the design space is useful to solve single-objective unconstrained optimization problems [1], but is is well known that they do not perform very well with the presence of constraints [2, 3, 4]. Furthermore, in many industrial applications, multiple objectives are pursued together.

Here is proposed a new method called *PAMUC (Preferences Applied to MUltiobjectivity and Constraints)* which handles simultaneously with the constrained and multiobjective aspects of general optimization problems.

91

T. Burczyński and A. Osyczka (eds),
IUTAM Symposium on Evolutionary Methods in Mechanics, 91–100.

2.   MULTIOBJECTIVE OPTIMIZATION IN EA

2.1   Definitions

This paragraph will describe basic concepts required to understand multiobjectivity, followed by an outline of the two families of techniques generally used to solve multiobjective problems.

General (unconstrained) multiobjective optimization problems can be written as follows :

$$\min_{x} \mathbf{f(x)} \tag{1}$$

where $\mathbf{f(x)}^{T} = [\ f_1(x)\ \ f_2(x)\ \dots\ f_m(x)\ ]$. The concept of Pareto optimum constitutes a major key to take multiobjectivity into account [5]. By definition, $\mathbf{x}^*$ is a Pareto optimum if there exists no other $\mathbf{x}$ (in the search space) such that, for at least one j, the following relation is satisfied :

$$f_j(\mathbf{x}) < f_j(\mathbf{x}^*) \tag{2}$$

In Pareto-based methods, the first step consists in drawing up the shape of the Pareto front. There are numerous methods to solve this problem : VEGA (Vector Evaluated Genetic Algorithm), MOGA (Multiple Objective Genetic Algorithm), ... [6]. Then, the user can make a choice among all the Pareto solutions.

Though this approach can give very interesting information about the design space, its main limitation lies in the fact that the determination of the Pareto front becomes a tedious task when the size of the problem increases. Consequently, it is often more advisable to use aggregating methods, where the user's preferences are taken into account since the very start of the process. Weights concerning the objectives can be affected thanks to a multi-criteria decision aid method, as PROMETHEE II. Its main features are described below.

2.2   PROMETHEE II

PROMETHEE II (*Preference Ranking Organisation METHod for Enrichment Evaluations II*) was incorporated for the first time in an EA by Rekiek, in order to optimize an assembly line design [7, 8]. Due to Brans and Mareschal [9], it provides a better interpretation of the weights thanks to a particular scaling of the objective functions. Furthermore, each individual is compared to all the other solutions of the population in order to build its fitness function. Here are the outlines of PROMETHEE II :

1. For each objective f_i a preference function is created, which compares any couple (a,b) of individuals. The parameters p_i and q_i must be defined by the user ($p_i = 1$ and $q_i = 0.01$ have provided good results in most cases).

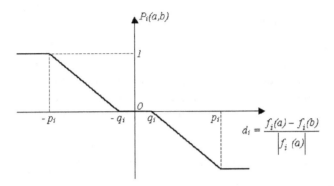

Figure 1. Preference function Pi(a,b) (for a minimization problem)

2. Then, the user affects a weight w_i to each objective function, in order to build the preference index of a over b :

$$(a,b) = \sum_{i=1}^{m} w_i P_i(a,b) \qquad (3)$$

3. Finally, to compare a solution a with the N-1 other solutions of a set E, the preference flow ! (a) is computed as follows :

$$!(a) = \frac{1}{N \square 1} \cdot \sum_{\substack{b \square E \\ b \square a}} (a,b) \qquad (4)$$

Consequently, the multiobjective problem is transformed into a single-objective problem with the preference flow ! acting like the fitness function. Then, to take the constraints into account, a new approach is proposed in the next paragraph.

3.   HOW TO HANDLE WITH CONSTRAINTS ?

The aim of this new method called PAMUC (*Preferences Applied to MUltiobjectivity and Constraints*) is to solve constrained multiobjective problems. The main idea is to use PROMETHEE II with m+1 objectives (the m objective functions and one more related to the satisfaction of the constraints). This latter function is estimated by :

$$f^{(m+1)}(\mathbf{x}) = \sum_{g_j < 0} \frac{|g_j(\mathbf{x})|}{k_j} \tag{5}$$

where $g_j(\mathbf{x})$ are inequality constraints (which are violated when $g_j(\mathbf{x}) < 0$) and k_j are scaling factors. The weights are initially chosen for the m objectives (w_i^{user} ; i = 1, ..., m). Then, at each generation t, they are computed as follows :

$$w_i^t = w_i^{user}.(1 - w_{m+1}^t) \quad \text{for } i = 1, ..., m \tag{6}$$

$$w_{m+1}^t = 1 - RF \tag{7}$$

where RF is the rate of the population which satisfies all the constraints and $w_{m+1} = 1$ at the first generation. The weights are thus adaptive : when the number of feasible individuals is low, the relative importance given to the objective m+1 (satisfaction of the constraints) is high ; but as the process goes on, a growing part of the population will satisfy the constraints, which will lead to a decrease of w_{m+1}.

4.   NUMERICAL RESULTS

The PAMUC method was incorporated in an EA programmed in Matlab. After validation on usual test functions, it has been implemented in Boss Quattro (Samtech), a commercial software for parametrical studies and optimization. Three examples will be presented, as described below :

Table 1. Description of the three examples presented

Optimization problem	Nb of variables	Nb of constraints	Nb of objectives
Himmelblau [3]	5	6	1
Unconstrained multiobjective problem (Cvetković [13])	2	0	2
Osyzcka [5]	2	2	2

4.1   Example 1

The first example was proposed by Himmelblau and used several times as a benchmark to test penalty-based techniques [3]. The objective function to be minimized is the following :

$$f(\mathbf{x}) = 5.3578547x_3{}^2 + 0.8356891x_1.x_5 + 37.293239.x_1 - 40792.141 \quad (8)$$

subject to 6 inequality constraints :

$$0 \,\square\, g_1(\mathbf{x}) \,\square\, 92 \,;\, 90 \,\square\, g_2(\mathbf{x}) \,\square\, 110 \,;\, 20 \,\square\, g_3(\mathbf{x}) \,\square\, 25 \qquad (9)$$

with the constraints $g_j(\mathbf{x})$ and the bounds of the variables defined as :

$$g_1(\mathbf{x}) = 85.334407 + 0.0056858x_2x_5 + 0.00026x_1x_4 - 0.0022053x_3x_5 \quad (10)$$

$$g_2(\mathbf{x}) = 80.51249 + 0.0071317x_2x_5 + 0.0029955x_1x_2 + 0.0021813x_3{}^2 (11)$$

$$g_3(\mathbf{x}) = 9.30096 + 0.0047026x_3x_5 + 0.0012547x_1x_3 + 0.0019085x_3x_4 (12)$$

$$78 \,\square\, x_1 \,\square\, 102 \,;\, 33 \,\square\, x_2 \,\square\, 45 \text{ and } 27 \,\square\, x_i \,\square\, 45 \text{ for } i = 3, \ldots, 5 \qquad (13)$$

The PAMUC method was implemented in an evolutionary algorithm programmed in Matlab, with a binary representation of the variables. Tournament selection scheme and 2-point crossover have been applied. The parameters and the results of the EA are summarized in Table 2, compared to results obtained with other constraint-handling methods (for more details, see [3]).

The PAMUC method provided good results with the same number of fitness function evaluations, compared to penalty-based methods, while the MGA (multi-objective genetic algorithm), based on a non-dominance approach [3], gives better results. It can also be noted that the co-evolutionary penalty method proposed by Coello [12], which has furnished the best results so far for this problem, needs a considerably higher number of evaluations (900000).

Table 2. Comparison of constraint-handling techniques for the Himmelblau function

	Static penalty (Homaifar *et al.*)	Dynamic penalty (Joines & Houck)	MGA (Coello)	PAMUC
Nb of fitness fct evaluations	5000	5000	5000	5000
Population size	50	50	50	100
Nb of generations	100	100	100	50
Crossover rate	0.8	0.8	self-adapted	0.8
Mutation rate	0.005	0.005	self-adapted	0.005
30 runs :				
Best solution	-30790	-30903	-31006	-30948
Mean	-30446	-30539	-30863	-30596
Standard deviation	226	200	73	206

The Figure 2 represents the fitness of the best feasible individual at each generation. A constant decrease of the objective function towards the optimum can be observed.

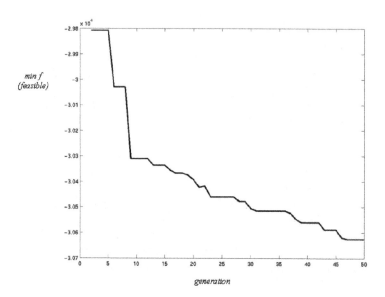

Figure 2. Evolution of fitness of the best feasible solution

4.2   **Example 2**

This example is an unconstrained maximization problem with 2 objective functions :

$$f_1(x_1,x_2) = \sin(x_1^2 + x_2^2 - 1) \tag{15}$$

$$f_2(x_1,x_2) = \sin(x_1^2 + x_2^2 + 1) \tag{16}$$

with x_1, $x_2 \in [0, 3/4]$. Three methods implemented in Boss Quattro (Samtech) have been used to solve this problem :

1. Multiple Objective Genetic Algorithm (Fonseca and Fleming, [6]) ;
2. Nondominated Sorting Genetic Algorithm (Srinivas and Deb, [6]) ;
3. PAMUC.

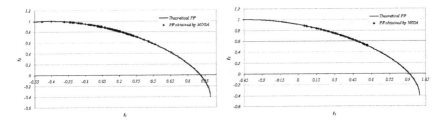

Figure 3. Theoretical and experimental Pareto Front (PF) by MOGA and NSGA techniques

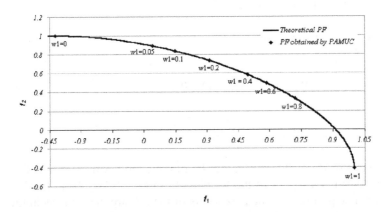

Figure 4. Theoretical and experimental PF by PAMUC

Real-coded variables were used, with the size of the population equal to 50 (in MOGA and NSGA) and to 30 (in PAMUC), and the maximum number of generations equal to 20. In order to draw the shape of the Pareto front in the PAMUC method, the weights affected to the objective functions have been modified (from $[w_1=1 ; w_2=0]$ to $[w_1=0 ; w_2=1]$). The Figures 3 and 4 show that although MOGA and NSGA provide a dense drawing of the Pareto front, the PAMUC method can give an interesting idea of its global shape, while NSGA for instance seems to converge towards the centre of the Pareto front.

It should be noted that in PAMUC, each point has to be computed by one entire run of the EA, while MOGA and NSGA techniques are designed to draw the entire shape of the Pareto front in a single run. Of course, PAMUC is not a Pareto-based method. Consequently, the aim of this example (as well as the following one) is not to show its efficiency in the drawing of Pareto fronts, but only to illustrate its ability to take correctly into account the user's preferences.

4.3   Example 3

The last problem presented in this paper is due to Osyczka [5], and is characterized by a concave shape of the Pareto front :

$$\max f_1(x_1,x_2) = x_1 + x_2^2 \tag{17}$$

$$\max f_2(x_1,x_2) = x_1^2 + x_2 \tag{18}$$

subject to 2 inequality constraints :

$$g_1(x_1,x_2) = 12 - x_1 - x_2 > 0 \tag{19}$$

$$g_2(x_1,x_2) = x_1^2 + 10x_1 - x_2^2 + 16x_2 - 80 > 0 \tag{20}$$

with : $2 \square x_1 \square 7$ and $5 \square x_2 \square 10$ (21)

The results of PAMUC are obtained by running the EA five times with a set of predetermined values of the weights (from $[w_1=1 ; w_2=0]$ to $[w_1=0 ; w_2=1]$ with a step of 0.25). The parameters of the EA are the same as in the previous example (size of the population = 30, number of generations = 20).

The experimental results (represented by black squares at Figure 5) are exactly located on the Pareto front.

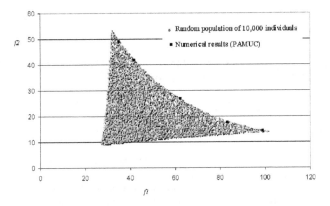

Figure 5. Random population of 10,000 individuals and results by PAMUC

4.4   Remarks

These test cases have provided good results of the PAMUC method to solve multiobjective problem with preferences as well as to handle with the constraints. An important advantage of the PAMUC method lies in the fact that it needs no particular tuning of the parameters (except the traditional parameters of the EA, as the size of the population, etc), while penalty-based methods usually requires a trial-and-error procedure before finding the best parameters values.

5.   CONCLUSIONS

In this paper, an original approach (PAMUC) was proposed to deal simultaneously with multiobjectivity and constraints in EA. The idea is to create a new objective function which takes the constraints into account. Then, an aggregating method, PROMETHEE II, is used in order to rank the individuals of the population at each generation. After a first implementation in Matlab, PAMUC was incorporated into a commercial software for parametrical studies and optimization, Boss Quattro (Samtech). It has provided good results for benchmarks, and will be applied shortly to an industrial application : the purge valve (designed by Techspace Aero) of the Vinci engine, from launcher Ariane 5. Future paths of research will include the use of expert rules, which are generally preponderant in the designers' choices.

ACKNOWLEGEMENTS

This work is supported by the Région Wallonne and by Samtech S.A.. Support from Techspace Aero is also acknowledged.

References

[1] D.E. Goldberg, Genetic Algorithms in Search, Optimization and Machine Learning, Addison-Wesley, Reading, MA (1989).

[2] Z. Michalewicz, D. Dasgupta, R.G. Le Riche, M. Schoenauer, Evolutionary algorithms for constrained engineering problems, Computers & Industrial Engineering Journal, Vol.30, No.2, pp. 851-870 (1996).

[3] C.A.C. Coello, Theoretical and numerical constraint-handling techniques used with evolutionary algorithms : a survey of the state of the art, Computer Methods in Applied Mechanics and Engineering, 191, pp. 1245-1287 (2002).

[4] A.H.C. van Kampen, C.S. Strom, and L.M.C. Buydens, Lethalization, Penalty and Repair Functions for Constraint Handling in the Genetic Algorithm Methodology, Chemometrics and Intelligent Laboratory Systems, pp. 55-68 (1996).

[5] D.A. Van Veldhuizen & G.B. Lamont, Multiobjective Evolutionary Algorithm Research : A History and Analysis, Graduate School of Engineering, Air Force Institute of Technology, Wright-Patterson AFB (1998).

[6] C.A.C. Coello, A Comprehensive Survey of Evolutionary-Based Multiobjective Optimization Techniques, Knowledge and Inform. Syst., An Intern. Journ., 1(3), pp. 269-308 (1999).

[7] B. Rekiek, Assembly Line Design : Multiple Objective Grouping Evolutionary Algorithm and the Balancing of Mixed-Model Hybrid Assembly Line, PhD thesis, Université Libre de Bruxelles (2001).

[8] P. De Lit, P. Latinne, B. Rekiek, A. Delchambre, Assembly planning with an ordering evolutionary algorithm, Intern. Journ. of Product. Research, vol. 39, n° 16, pp. 3623-3640 (2001).

[9] J.-P. Brans, B. Mareschal, The PROMCALC and GAIA decision support system for multi-criteria decision aid, Decision Support Systems, North Holland, vol. 12, pp. 297-310 (1994).

[10] J.A. Joines and C.R. Houck, On the use of non-stationary penalty functions to solve nonlinear constrained optimization problems with GAs, Proceedings of the First IEEE International Conference on Evolutionary Computation, pp. 579-584, IEEE Press (1994).

[11] Deb, An efficient constraint handling method for genetic algorithms, Computer Methods in Applied Mechanics and Engineering, 186, pp. 311-338 (2000).

[12] C.A.C. Coello, Use of a self-adaptive penalty approach for engineering optimization problems, Comput. Ind. 41 (2), pp. 113-127 (2000).

[13] D. Cvetković, Evolutionary Multi-Objective Decision Support Systems for Conceptual Design, PhD Thesis, School of Computing, Faculty of Technology, University of Plymouth (2000).

A COMPARATIVE ANALYSIS OF "CONTROLLED ELITISM" IN THE NSGA-II APPLIED TO FRAME OPTIMIZATION

D. Greiner, G. Winter, J.M. Emperador and B. Galván
Intelligent Systems and Numerical Applications in Engineering Institute (IUSIANI),
Evolutionary Computation and Applications Division (CEANI), Las Palmas de Gran Canaria
University, Spain

Abstract: Controlled elitism in NSGAII is analysed, applied to a multiobjective frame
optimisation problem with discrete modelling. Influence of various mutation
rates is considered. A double minimization is handled: constrained weight and
number of different cross section types. The comparative statistical results of
the test case show a convergence study during evolution by means of certain
metrics that measure front coverage and distance to the optimal front.

Keywords: Multiobjective Optimisation, Structural Optimisation, NSGA-II, Elitism.

1.   INTRODUCTION

The evolutionary multicriterion methods have acquired importance, due to the lack of disadvantages that the classic methods have, being applied to a wide range of problems that were restricted to the classical [1]. The article describes the NSGAII and controlled elitism [2]. Later, the problem and test case are described; at last results via some metrics, analysis and conclusions.

2.   NSGA-II AND CONTROLLED ELITISM (CE)

The NSGA-II [3] is originated from the NSGA including elitism and avoiding the parameter dependence of sharing factor. It also introduces a faster algorithm to sort the population. It evaluates a crowding distance that

101

T. Burczyński and A. Osyczka (eds),
IUTAM Symposium on Evolutionary Methods in Mechanics, 101–110.

considers the size of the largest cuboid enclosing each population member without including any other. This parameter-less magnitude is used to keep diversity in the population. Selection is performed by a tournament selection, that prioritises the front domination order and the crowding distance.

CE offers a quantitative control over the algorithm selection pressure. Its activity is focused in two factors: Smoothes the elitism of the NSGA-II, and limits the max-number of individuals belonging to each front. Elitism in the NSGA-II is in two phases. In a first stage, the next-generation population is selected from the best individuals (based on the non-dominance criterium) belonging to the parent and offspring population of the present generation. Then, a tournament selection is done to decide which individuals are crossed and muted: it is a second elitism stage, increasing the selection pressure. With CE, this second stage is deleted, replacing the tournament selection with a random procedure. Moreover, it limits the max-number of individuals belonging to each front by a geometric decreasing function governed by the reduction rate r. So, we impose the existence of a certain number of fronts in the population, avoiding a possible fast convergence to only non-dominated individuals. Being K the number of desired fronts, N the total number of individuals in the population, and r the selected reduction factor, $(r < 1)$, then the number of individuals n_i in each i front, is [2]:

$$n_i = N \frac{1-r}{1-r^K} r^{i-1}$$

(1)

3.   THE MULTIOBJECTIVE FRAME PROBLEM

The frame problem consists in minimizing two criteria [4][5]:

1. The **constrained weight**, optimising the acquisition cost of the metallic frame raw material. It is obtained through the direct stiffness method and the following constraints are applied.

a) *Stresses of the bars*, where the comparing stress takes into account the axial and shearing stresses by the shear effort, and the bending effort.

b) *Compressive slenderness limit*, for each bar whose buckling effect is considered (the Spanish norm has a limit of 200).

c) *Displacements of joints or middle points* of bars.

$$\sigma_{co} - \sigma_{lim} \leq 0$$

(2)

$$\lambda - \lambda_{lim} \leq 0$$

(3)

$$u_{co} \square u_{lim} \square 0 \tag{4}$$

Considering the variables (refered to bar *i*):
A_i = area of the section type; $\square$ = specific weight; l_i = length; k = constant that regulates the equivalence between weight and restriction; $viol_j$ = for each violated restriction j, is the quotient between the violated restriction value (stress, displacement or slenderness) and its reference limit.
With these considerations, the fitness function *constrained weight* results:

$$Fitness = \left[\sum_{i=1}^{Nbars} A_i \,\square\square_i \,\square_i \right] [1 + k \,\square \sum_{j=1}^{Nviols} (viol_j \,\square 1)] \tag{5}$$

2. The **number of different cross section types**, due to constructive constraints, of importance in structures with a high number of bars. It helps to a better quality control during the execution of the building site. The algorithm needs (n-1) comparisons in the most favourable case:

```
nudicrosecty = 0
for(i=0; i<barnumber; i++){
    loopend = FALSE;
    for(j=(i+1);j< barnumber && not(loopend);j++){
        if( sectiontype(j) = sectiontype(i) ){
            nudicrosecty = nudicrosecty + 1;
            loopend = TRUE;
    } } }
nudicrosecty = barnumber - nudicrosecty;
```

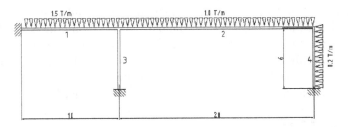

Figure 1. Structural Test Case

4.   TEST CASE

The test case (fig. 1) is based on a problem taken from Hernández [6]. The spot lengths are in meters, and there is a max-displacement constraint in bar 2 middle point, of length/300. It has been considered buckling effect and own gravitational load. Fig 2 shows the Pareto front (x-axis: frame weight; y-axis: number of different cross-section types, belonging to IPE class).

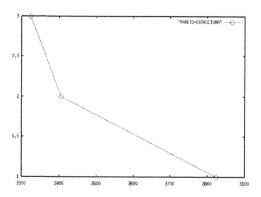

Figure 2. Pareto Front of the structural test case

5.   EXPERIMENTAL RESULTS

5.1   General Parameters

We consider 30 independent runs of each algorithm and a population size of 100. Tested algorithms are the standard NSGAII and the NSGAII with CE. To investigate independently the influence in the exploitation–exploration equilibrium, mutation rate values (0%,0.4%,3%) and reduction factor r (0.0,0.5) have been considered. In nine studied combinations, uniform crossover, gray code and $K=5$ are used.

5.2   Used Metrics

Two metrics are considered; the Metric 1 is the M1* metric of Zitzler [9], representative of the approximation to the optimal Pareto front:

$$M1*(U) = \frac{1}{|U|} \sum_{u \square U} \min \left\{ \|u \square y\| \mid y \square Y_p \right\}$$

(6)

,where U = f(A) □ Y.

The Metric 2 handles with the coverage of the front. Being this a discrete domain problem, we have considered the number of different cross section types covering as Metric 2. Its reference value is three.

5.3   Figures

The average results of the thirty executions for each metric are grouped in two blocks. The first (figs. 3 to 8), shows the results by algorithm and comparing the different studied mutation rates. In this figures, the symbols diamond, cross and square correspond to mutation rates of 0%, 0.4% and 3% respectively. A second (figs. 9 to 14), shows the results by mutation rate and comparing the algorithms. The symbols diamond, cross and square correspond to NSGAII algorithms: standard, CE & r=0.5; and CE & r=0.0.

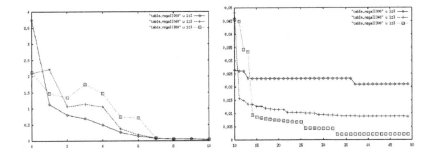

Figure 3. Metric 1 in the NSGAII from generation 0 to 10 and from 10 to 50th

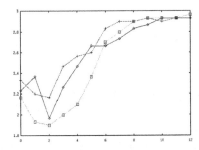

Figure 4. Metric 2 in the NSGAII from generation 0 to 12th

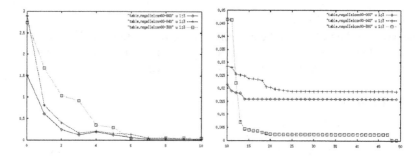

Figure 5. Metric 1 in the NSGAII-CE and r=0.0 from generation 0 to 10 and from 10 to 50th

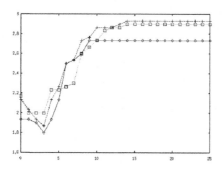

Figure 6. Metric 2 in the NSGAII-CE and r=0.0 from generation 0 to 25th

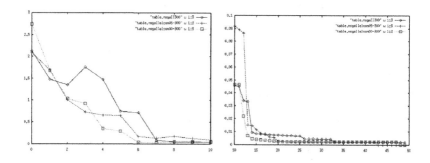

Figure 7. Metric 1 in the NSGAII-CE and r=0.5 from generation 0 to 10 and from 10 to 50th

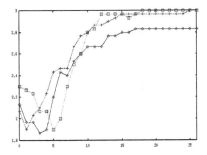

Figure 8. Metric 2 in the NSGAII-CE and r=0.5 from generation 0 to 26th

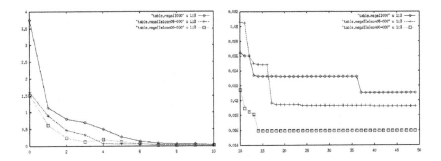

Figure 9. Metric 2 in the NSGAII-CE and r=0.5 from generation 0 to 26th

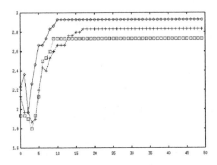

Figure 10. Metric 2 with mutation rate 0% from generation 0 to 50th

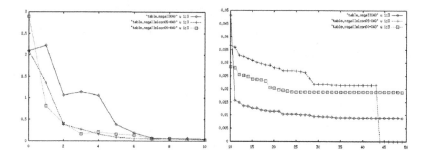

Figure 11. Metric 1 with mutation rate 0.4% from generation 0 to 10 and from 10 to 50th

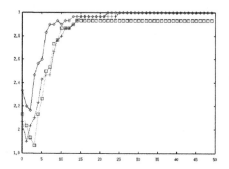

Figure 12. Metric 2 with mutation rate 0.4% from generation 0 to 50th

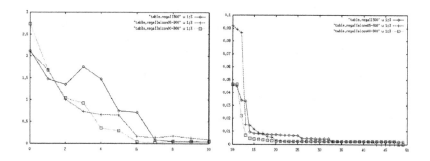

Figure 13. Metric 1 with mutation rate 3% from generation 0 to 10 and from 10 to 50th

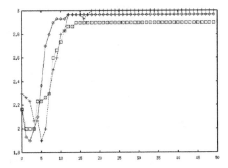

Figure 14. Metric 2 with mutation rate 3% from generation 0 to 50th

6.   ANALYSIS OF RESULTS

6.1   Algorithms

From observation of Metric 1 in figs. 3 to 8, it can be seen a common behaviour in the three tested algorithms. In initial generations, the algorithm with higher mutation rate experiments a slower convergence to the Pareto front, but about the twelfth generation the process becomes inverse; the evolution in the mutation-less algorithm stagnates and the clear winner is the higher mutation rate. The algorithm with 0.4% mutation rate is intermediate. So, the exploration that gives mutation in the final stages is advantageous; but its lack, in early stages accelerates convergence (giving more intensive exploitation). In case of Metric 2, we can also observe a similar behaviour of the three algorithms. It diminishes initially, due to the progressive finding of non-dominating individuals in the initial stages, and not a complete front. Later, when more members are included, the front grows. The mutation-less case stagnates and offers a lower average coverage, being advantageous the exploration of mutation. This is relevant in case of algorithms with CE: mutation and CE can be complementary, interacting at same time positively.

6.2   Mutation Rates

In figs. 9 to 14, the different algorithms are compared with the same mutation rate. Figures 9 and 10, correspond to the lack of mutation operator. In this case, population diversity exclusively depends on the CE. We can observe, that standard NSGA-II shows the slowest convergence in the initial stages, and also in the final ones. Both algorithms with CE have a faster behaviour. In figure 13, (3% of mutation rate in Metric 1), we can see a similar convergence of the three algorithms in the last generations, but a clear disadvantage of the standard NSGAII in the initial generations. This advantage is also present in the initial stages of the case of 0.4% of mutation rate. It is also appreciable that in the figures of Metric 2, the front coverage in the final stages is slightly lower for the case of higher mutation.

7.   CONCLUSIONS

A comparative analysis of the exploration-exploitation equilibrium in a frame multiobjective optimisation problem is discussed, taking into account both mutation and 'controlled elitism' (CE) in the NSGA-II. The obtained results show that the CE operator offers advantages in the convergence of the algorithm, being capable to cooperate with the mutation operator. The treated problem is solved satisfactorily, and the NSGA-II with CE configures as a competitive algorithm in discrete multiobjective frame optimization.

ACKNOWLEDGEMENTS

This investigation was funded by a FPU - Spanish Government Grant of The Ministry of Education and Culture - corresponding to David Greiner.

REFERENCES

[1] C. Coello Coello, "A Short Tutorial on Evolutionary Multiobjective Optimization", pp. 21-40, in Evolutionary Multi-Criterion Optimization, Springer, 2001.

[2] K. Deb, T. Goel, "Controlled Elitist Non-dominated Sorting Genetic Algorithms for Better Convergence", pp. 67-81, in Evolutionary Multi-Criterion Optimization, Springer, 2001.

[3] K. Deb, A. Pratap, S. Agrawal, T. Meyarivan, "A fast and elitist multiobjective genetic algorithm NSGAII", *IEEE Transactions on Evolutionary Computat.* 6(2),182-197, 2002.

[4] D. Greiner, G. Winter, J.M. Emperador, "Optimising Frame Structures by different strategies of GA", *Finite Elem. in Anays. and Design*, Elsevier, 37(5) pp.381-402, (2001).

[5] P. Hajela, C. Lin. 'Genetic search strategies in multi-criterion optimal design', *Structural Optimization*,4 pp 99 – 107 (1992).

[6] S. Hernández Ibáñez. 'Métodos de diseño óptimo de estructuras'. Colección Seinor. Colegio de Ingenieros de Caminos, Canales y Puertos (Madrid). (1990).

[7] R. Purshouse, P. Fleming, "Why use Elitism and Sharing in a Multiobjective Genetic Algorithm?", *GECCO-2002*, pp. 520-527, New York, Morgan Kaufmann Publishers.

[8] S. Rajeev, C.S. Krishnamoorthy. 'Discrete Optimization of Structures using Genetic Algorithms'. *Journal of Structural Engineering*, vol 118, nº 5, 1233-1250 (1992).

[9] E. Zitzler, Evolutionary Algorithms for Multiobjective Optimization: Methods and Applications, PhD Thesis, 1999.

IS-PAES: MULTIOBJECTIVE OPTIMIZATION WITH EFFICIENT CONSTRAINT HANDLING

Arturo Hernández Aguirre
Salvador Botello Rionda
Giovanni Lizárraga Lizárraga
Center for Research in Mathematics
Department of Computer Science, Guanajuato, Gto. 36240, MEXICO
artha,botello,giovanni@cimat.mx

Carlos Coello Coello
CINVESTAV-IPN, EE Dept., Computer Science Section
México, D.F. 07300, MEXICO
ccoello@cs.cinvestav.mx

Abstract: This paper introduces a new constraint-handling method called Inverted-Shrinkable PAES (IS-PAES), which focuses the search effort of an evolutionary algorithm on specific areas of the feasible region by shrinking the constrained space of single-objective optimization problems. IS-PAES uses an adaptive grid as the original PAES (Pareto Archived Evolution Strategy). However, IS-PAES does not have the serious scalability problems of the PAES. The viability of the proposed approach is validated with several examples taken from the standard evolutionary and engineering optimization literature.

Keywords: multiobjective optimization, constraint handling, PAES

1. INTRODUCTION

Evolutionary Algorithms (EAs) in general (i.e., genetic algorithms, evolution strategies and evolutionary programming) lack a mechanism able to bias efficiently the search towards the feasible region in constrained search spaces. Such a mechanism is highly desirable since most real-world problems have constraints which could be of any type (equality, inequality, linear and nonlinear). The success of EAs in global optimization has triggered a considerable amount of research regarding the development of mechanisms able to incorporate information about

111

T. Burczyński and A. Osyczka (eds),
IUTAM Symposium on Evolutionary Methods in Mechanics, 111–120.

the constraints of a problem into the fitness function of the EA used
to optimize it [4, 7]. So far, the most common approach adopted in
the evolutionary optimization literature to deal with constrained search
spaces is the use of penalty functions. When using a penalty function,
the amount of constraint violation is used to punish or "penalize" an
infeasible solution so that feasible solutions are favored by the selection
process. Despite the popularity of penalty functions, they have several
drawbacks from which the main one is that they require a careful fine
tuning of the penalty factors that indicates the degree of penalization to
be applied [9, 4]. Recently, some researchers have suggested the use of
multiobjective optimization concepts to handle constraints in EAs (see
for example [4]). This paper introduces a new approach that is based
on an evolution strategy that was originally proposed for multiobjective optimization: the Pareto Archived Evolution Strategy (PAES) [6].
Our approach can be use to handle constraints both of single- and multiobjective optimization problems and does not present the scalability
problems of the original PAES. The remainder of this paper is organized
as follows. Section 2 briefly describes the work related to our own. In
Section 3, we describe the main algorithm of IS-PAES. Section 4 provides a comparison of results and Section 5 draws our conclusions and
provides some paths of future research.

2.   RELATED WORK

Since our approach belongs to the group of techniques in which multiobjective optimization concepts are adopted to handle constraints, we
will briefly discuss some of the most relevant work done in this area.
The main idea of adopting multiobjective optimization concepts to handle constraints is to redefine the single-objective optimization of $f(\vec{x})$
as a multiobjective optimization problem in which we will have $m + 1$
objectives, where m is the total number of constraints. Then, we can
apply any multiobjective optimization technique [5] to the new vector
$\bar{v} = (f(\vec{x}), f_1(\vec{x}), \ldots, f_m(\vec{x}))$, where $f_1(\vec{x}), \ldots, f_m(\vec{x})$ are the original constraints of the problem. An ideal solution $\vec{x}$ would thus have $f_i(\vec{x})=0$ for
$1 \leq i \leq m$ and $f(\vec{x}) \leq f(\vec{y})$ for all feasible $\vec{y}$ (assuming minimization).

Based on this main idea, several approaches have proposed in the last
few years. Some of them use population-based techniques (e.g., [3]), and
others use Pareto ranking (e.g., [10]). However, all of these techniques
are normally more useful to approach the feasible region, but are not
as effective for reaching the global optimum of a problem. We argue
in this paper that the main reason for having this limitation has to do
with the focus of the search in traditional multiobjective optimization

algorithms. Rather than focusing the effort on finding good "trade-offs" (as in multiobjective optimization), we propose to focus the search in finding the boundary between the feasible and the infeasible regions and then concentrating the search effort on reaching the global optimum. Such is the nature of the algorithm proposed in this paper.

3.   IS-PAES ALGORITHM

IS-PAES has been implemented as an extension of the Pareto Archived Evolution Strategy (PAES) proposed by Knowles and Corne [6] for multiobjective optimization. PAES main feature is the use of an adaptive grid on which objective function space is located using a coordinate system. Such a grid is the diversity maintenance mechanism of PAES and its the main feature of this algorithm. The grid is created by bisecting k times the function space of dimension $d = g + 1$. The control of 2^{kd} grid cells means the allocation of a large amount of physical memory for even small problems. For instance, 10 functions and 5 bisections of the space produce 2^{50} cells. Thus, the first feature introduced in IS-PAES is the "inverted" part of the algorithm that deals with this space usage problem. IS-PAES's fitness function is mainly driven by a feasibility criterion. Global information carried by the individuals surrounding the feasible region is used to concentrate the search effort on smaller areas as the evolutionary process takes place. In consequence, the search space being explored is "shrinked" over time. Eventually, upon termination, the size of the search space being inspected will be very small and will contain the solution desired. The main algorithm of IS-PAES is shown in Figure 1.

The function **test(h,c,file)** determines if an individual can be added to the external memory or not. Here we introduce the following notation: $x_1 \square x_2$ means x_1 is located in a less populated region of the grid that x_2. The pseudo-code of this function is depicted in Figure 2.

3.1   Inverted "ownership"

IS-PAES handles the population *as part of* a grid location relationship, whereas PAES handles a grid location *contains* population relationship. In other words, PAES keeps a list of individuals on either grid location, but in IS-PAES either individual knows its position on the grid. Therefore, building a sorted list of the most dense populated areas of the grid only requires to sort the k elements of the external memory. In PAES, this procedure needs to inspect every location of the grid in order to produce an unsorted list, there after the list is sorted. The advantage of the inverted relationship is clear when the optimization problem has

maxsize: max size of file
c: current parent $\in X$ (decision variable space)
h:child of $c \in X$, a_h: individual in file that dominates h
a_d: individual in file dominated by h
current: current number of individuals in file
cnew: number of individuals generated thus far
current = 1; cnew=0; c = newindividual() ; add(c)
While cnew$\leq$MaxNew **do**
   h = mutate(c); cnew+ =1;
   if (c$\preceq$h) **then exit loop**
   else if (h$\prec$c) **then** { remove(c); add(g); c=h; }
   else if ($\exists$ $a_h \in$ file | $a_h \preceq$ h) **then exit loop**
   else if ($\exists$ $a_d \in$ file | h $\preceq$ a_d) **then**
      add(h); $\forall$ a_d { remove(a_d); current$-$ =1 }
   else test(h,c,file)
   if (cnew % g==0) **then** c = individual in less densely populated region
   if (cnew % r==0) **then** shrinkspace(file)
End While

Figure 1.   Main algorithm of IS-PAES

if (current < maxsize) **then** add(h)
   if (h $\square$ c) **then** c = h
else if ($\exists a_p \in$file | h $\square$ a_p) **then** { remove(a_p); add(h) }
if (h $\square$ c) **then** c = h;

Figure 2.   Pseudo-code of **test(h,c,file)**

many functions (more than 10), and/or the granularity of the grid is fine, for in this case only IS-PAES is able to deal with any number of functions and granularity level.

3.2   Shrinking the objective space

Shrinkspace(file) is the most important function of IS-PAES since its task is the reduction of the search space. The pseudo-code of **Shrinkspace(file)** is shown in Figure 3. The function **select(file)** returns a list whose elements are the best individuals found in *file*. The size of the list

$\underline{x}_{pob}$: vector containing the smallest value of either $x_i \in X$
$\overline{x}_{pob}$: vector containing the largest value of either $x_i \in X$
select(file); getMinMax(file, $\underline{x}_{pob}$, $\overline{x}_{pob}$)
trim($\underline{x}_{pob}$, $\overline{x}_{pob}$)
adjustparameters(file);

Figure 3. Pseudo-code of **Shrinkspace(file)**

is 15% of *maxsize*. Since individuals could be feasible, infeasible or only partially feasible, the best individuals are chosen from the *file* sorted by "objective-function" then by "constraint#1", then by "constraint#2", and so on. The function **getMinMax(file)** takes this list and finds the extreme values of the decision variables represented by those individuals. Thus, the vectors $\underline{x}_{pob}$ and $\overline{x}_{pob}$ are found. Function **trim($\underline{x}_{pob}$, $\overline{x}_{pob}$)** shrinks the feasible space around the potential solutions enclosed in the hypervolume defined by the vectors $\underline{x}_{pob}$ and $\overline{x}_{pob}$. Thus, the function **trim()** (see Figure 4) determines the new boundaries for the decision variables.

n: size of decision vector;
$\overline{x}_i$: actual upper bound of the i_{th} decision variable
$\underline{x}_i$: actual lower bound of the i_{th} decision variable
$\overline{x}_{pob,i}$: upper bound of i_{th} decision variable in population
$\underline{x}_{pob,i}$: lower bound of i_{th} decision variable in population
$\forall i : i \in \{ 1, \ldots, n \}$
$\quad slack_i = 0.05 \times (\overline{x}_{pob,i} - \underline{x}_{pob,i})$
$\quad width\_pob_i = \overline{x}_{pob,i} - \underline{x}_{pob,i}; \; width_i^t = \overline{x}_i^t - \underline{x}_i^t$
$\quad deltaMin_i = \frac{\beta * width_i^t - width\_pob_i}{2}$
$\quad delta_i = \max(slack_i, deltaMin_i);$
$\quad \overline{x}_i^{t+1} = \overline{x}_{pob,i} + delta_i; \; \underline{x}_i^{t+1} = \underline{x}_{pob,i} - delta_i;$
$\quad$ **if** $(\overline{x}_i^{t+1} > \overline{x}_{original,i})$ **then**
$\quad\quad \underline{x}_i^{t+1} - = \overline{x}_i^{t+1} - \overline{x}_{original,i}; \; \overline{x}_i^{t+1} = \overline{x}_{original,i};$
$\quad$ **if** $(\underline{x}_i^{t+1} < \underline{x}_{original,i})$ **then** $\overline{x}_i^{t+1} + = \underline{x}_{original,i} - \underline{x}_i^{t+1};$
$\quad\quad \underline{x}_i^{t+1} = \underline{x}_{original,i};$
$\quad$ **if** $(\overline{x}^{t+1} > \overline{x}_{original,i})$ **then** $\overline{x}_i^{t+1} = \overline{x}_{original,i};$

Figure 4. Pseudo-code of **trim**

The value of β is the percentage by which the boundary values of either $x_i \in X$ must be reduced such that the resulting hypervolume is a fraction α of its initial value. In our experiments, $\alpha = 0.90$ worked well in all cases. Clearly, α controls the shrinking speed, hence the algorithm is sensitive to this parameter and it can prevent it from finding the optimum solution if small values are chosen. In our experiments, values in the range [85%,95%] were tested with no visible effect in the performance. Of course, α values near to 100% slow down the convergence speed. The last step of **shrinkspace()** is a call to **adjustparameters(file)**. The goal is to re-start the control variable σ using: $\sigma_i = (\overline{x}_i - \underline{x}_i)/\sqrt{n}$ $i \in (1, \ldots, n)$ This expression is also used during the generation of the initial population. In that case, the upper and lower bounds take the initial values of the search space indicated by the problem. The variation of the mutation probability follows the exponential behavior suggested by Bäck [1].

4. COMPARISON OF RESULTS

We have validated our approach with several problems used as a benchmark for evolutionary algorithms (see [7]) and with several engineering optimization problems taken from the standard literature. In the first case, our results are compared against a technique called "stochastic ranking" which is representative of the state-of-the-art in constrained evolutionary optimization (see [8] for details). Regarding the engineering optimization problems, we compared results with respect to those previously reported in the literature.

4.1 Test problems used with EAs

The following parameters were adopted in this case: $maxsize = 200$, $bestindividuals = 15\%$, $slack = 0.05$, $r = 400$, $MaxNew = 350000$.

1 Problem g06

$$\text{Minimize } F(\boldsymbol{x}) = (x_1 - 10)^3 + (x_2 - 20)^3 \tag{1}$$

subject to:

$$g_1(\boldsymbol{x}) = -(x_1 - 5)^2 - (x_2 - 5)^2 + 100 \leq 0$$
$$g_2(\boldsymbol{x}) = (x_1 - 6)^2 + (x_2 - 5)^2 - 82.81 \leq 0 \tag{2}$$

where $13 \leq x_1 \leq 100$ y $0 \leq x_2 \leq 100$. The global optimum is located at $\boldsymbol{x}^* = \{14.095, 0.84296\}$, with $F(\boldsymbol{x}^*) = -6961.81388$. Both constraint are active. The best solution found by IS-PAES is: $\boldsymbol{x} = \{14.0950000092\ 0.842960808844\ \}$, with $F(\boldsymbol{x}) = -6961.813854$.

Note in Table 1 how IS-PAES had a better average result and a lower standard deviation than R&Y.

Measure	IS-PAES	R&Y
Best	-6961.814	-6961.814
Worst	-6961.810	-6350.262
Average	-6961.813	-6875.940
Std. Deviation	8.5E-05	1.6E+02
Median	-6961.814	-6961.814
Feasible solutions	30	30

Table 1. Comparison of results for problem g06

2 Problem g08

$$\text{Minimize} F(\boldsymbol{x}) = \frac{\sin^3(2\pi x_1)\sin(2\pi x_2)}{x_1^3(x_1 + x_2)} \tag{3}$$

subject to:

$$g_1(\boldsymbol{x}) = x_1^2 - x_2 + 1 \le 0$$
$$g_2(\boldsymbol{x}) = 1 - x_1 + (x_2 - 4)^2 \le 0 \tag{4}$$

where $0 \le x_1 \le 10$ y $0 \le x_2 \le 10$. The global optimum is located at $\boldsymbol{x}^* = \{1.2279713, 4.2453733\}$, with $F(\boldsymbol{x}^*) = 0.095825$. The best solution found by IS-PAES was: $\boldsymbol{x} = \{1.227971353, 4.245373368\}$, with $F(\boldsymbol{x}) = -0.095825041$. The performance of both algorithms is similar for this problem. Results are shown in Table 2.

Measure	IS-PAES	R&Y
Best	-0.095825	-0.095825
Worst	-0.095825	-0.095825
Average	-0.095825	-0.095825
Std. Deviation	0.0	2.6E-17
Median	-0.095825	-0.095825
Feasible solutions	30	30

Table 2. Comparison of results for problem g08

4.2   Optimization of a 49-bar plane truss

The last problem is the optimization of a 49-bar plane truss. The goal is to find the cross-sectional area of each member of the truss, such that the overall weight is minimized, subject to stress and displacement constraints. The weight of the truss is given by $F(\boldsymbol{x}) = \sum_{j=1}^{49} \gamma A_j L_j$, where A_j is the cross-sectional area of the j_{th} member, L_j is the corresponding length of the bar, and γ is the volumetric density of the material. We used a catalog of *Altos Hornos de México, S.A.*, with 65 entries for the cross-sectional areas available for the design. Other relevant data are the following: Young modulus $= 2.1 \cdot 10^6 kg/cm^3$, maximum allowable stress$= 3500.00 kg/cm^2$, $\gamma = 7.4250 \cdot 10^{-3}$, and a horizontal load of 4994.00 kg applied to the nodes: 3, 5, 7, 9, 12, 14, 16, 19, 21, 23, 25 y 27. We solved this problem for two cases:

Case 1. Stress constraints only. Maximum allowable stress $=$ $3500.00 kg/cm^2$ A total of 49 constraints, thus 50 objective functions.

Case 2. Stress and displacement constraints. Maximum allowable stress $= 3500.00 kg/cm^2$, maximum displacement per node $= 10cm$ A total of 72 constraints, thus 73 objective functions.

Case 3. Real design problem. The design problem considers traction and compression stress on the bars, as well as the proper weight. Maximum allowable stress $= 3500.00 kg/cm^2$, maximum displacement per node $= 10cm$. A total of 72 constraints, thus 73 objective functions. The average result of 30 runs for each case are shown in Tables 3, 4 and 5. We compare IS-PAES with previous results reported by Botello [2] (SA: Simulated Annealing, GA50: Genetic Algorithm with a population of 50, and GSSA: General Stochastic Search Algorithm with populations of 50 and 5).

Algorithm	Average Weight (Kg)
IS-PAES	610
SA	627
GA50	649
GSSA50	619
GSSA5	625

Table 3. Comparison of different algorithms on the 49-bar struss, case 1

5. CONCLUSIONS AND FUTURE WORK

We have introduced a constraint-handling approach that combines multiobjective optimization concepts with an efficient reduction mechanism of the search space and a secondary population. We have shown how our approach overcomes the scalability problem of the original PAES from which it was derived, and we also showed that the approach is highly competitive with respect to the state-of-the-art in the area. As

Algorithm	Average Weight (Kg)
IS-PAES	725
SA	737
GA50	817
GSSA50	748
GSSA5	769

Table 4. Comparison of different algorithms on the 49-bar struss, case 2

Algorithm	Average Weight (Kg)
IS-PAES	2603
SA	2724
GA50	2784
GSSA50	2570
GSSA5	2716

Table 5. Comparison of different algorithms on the 49-bar struss, case 3

part of our future work, we want to refine the mechanism adopted for reducing the search space being explored, since in our current version of the algorithm, premature convergence may occur in some cases. The elimination of the parameters required by our approach is another goal of our current research. Finally, we also intend to couple the mechanisms proposed in this paper to other multiobjective evolutionary algorithms.

ACKNOWLEDGMENTS

The first author acknowledges partial support from CONCyTEG project No. 01-02-202-111 and CONACyT No. I-39324-A. The second author acknowledges support from CONACyT project No. 34575-A. The last author acknowledges support from CONACyT project No. NSF-CONACyT 32999-A.

References

[1] Thomas Bäck, *Evolutionary Algorithms in Theory and Practice*, COxford University Press, 1996.

[2] Salvador Botello and José Luis Marroquín and et al., *Solving Structural Optimization problems with GAs and Simulated Annealing*, International Journal of Numerical Methods in Engineering, vol. 45, pp. 1-16, 1999.

[3] Carlos A. Coello Coello, *Treating constraints as objectives for single-objective evolutionary optimization*, Engineering Optimization, vol. 32, pp. 275-308, 2000.

[4]   Carlos A. Coello Coello, *Theoretical and Numerical Constraint Handling Techniques used with Evolutionary Algorithms: A Survey of the State of the Art,* Computer Methods in Applied Mechanics and Engineering, vol. 191, pp. 1245-1287, 2002.

[5]   Carlos A. Coello Coello and David A. Van Veldhuizen, *TEvolutionary Algorithms for Solving Multi-Objective Problems,* Kluwer Academic Publishers, 2002.

[6]   J.D. Knowles and D.W. Corne, *Approximating the Nondominated Front using the Pareto Archived Evolution Strategy,* Evolutionary Computation, vol. 2, pp. 149-172, 2000.

[7]   Zbigniew Michalewicz and Marc Schoenauer, *Evolutionary Algorithms for Constrained Parameter Optimization Problems,* Evolutionary Computation, vol. 4, pp. 1-32, 1996.

[8]   T.P. Runarsson and X. Yao, *Stochastic Ranking for Constrained Evolutionary Optimization,* IEEE Transactions on Evolutionary Computation, vol. 4, pp. 284-294, 2000.

[9]   Alice E. Smith and David W. Coit, *Constraint Handling Techniques—Penalty Functions,* in Handbook of Evolutionary Computation, ed.Thomas Bäck and David B. Fogel and Zbigniew Michalewicz, Oxford University Press and Institute of Physics Publishing, 1997.

[10]   Patrick D. Surry and Nicholas J. Radcliffe, *The COMOGA Method: Constrained Optimisation by Multiobjective Genetic Algorithms,* Control and Cybernetics, vol. 26, pp. 391-412, 1997.

OPTIMIZATION OF ALIGNED FIBER LAMINATE COMPOSITES

Design in the Presence of a Stress Concentration

Zhenning Hu[1], Guy M. Genin[1], Mark J. Jakiela[1], Alex M. Rubin[2]
[1]*Department of Mechanical Engineering, Washington University, St. Louis, MO 63130, USA*
[2]*The Boeing Company, St. Louis, MO, 63166 USA*

Abstract: A fundamental problem in the design of laminate composites is strength degradation in the presence of stress concentrating features such as flaws and attachment holes. This paper presents a scheme for optimizing material design so as to maximize strength in the presence of flaws. This optimization occurs over a landscape littered with local maxima, which is reliably traversed by a heuristic approach such as a genetic algorithm. A simple example is presented which highlights the benefits and drawbacks of considering flaw tolerance in laminate design.

Keywords: fiber reinforced composites, genetic algorithms, stress concentrations

1.   INTRODUCTION

Fiber-reinforced laminate composites are in widespread use because of their superior ratios of strength and stiffness to weight. These materials are constructed by embedding strong, stiff, long fibers (e.g., Carbon fibers) in a weaker matrix (e.g., epoxy) to form orthotropic laminae like that shown in Figure 1a. The resulting orthotropic variations in strength and stiffness can be employed to the advantage of the designer, but can also lead to amplification of stress concentration factors around stress-concentrating defects.

121

T. Burczyński and A. Osyczka (eds),
IUTAM Symposium on Evolutionary Methods in Mechanics, 121–129.
© *2004 Kluwer Academic Publishers.*

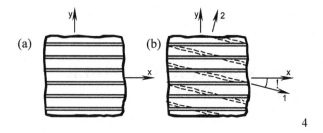

Figure 1. (a) Lamina, (b) laminate, with definition of !.

To highlight the severity of the amplification of stress concentration due to material anisotropy, the problem of stress concentration is illustrated in Figure 2 for the case of an infinite orthotropic plate, loaded in uniform tension, that contains a central circular hole. An analytical solution is available for this problem [4,10]. The plate studied has identical Young's moduli E in the directions parallel and perpendicular to the load; three values of Poisson's ration $\square$ are shown. As the shear modulus G for shearing in axes parallel and perpendicular to the loading direction decreases from the isotropic value $E/(2(1+\square))$, the maximum stress concentration factor along the hole boundary (maximum stress $\square_{\square}^{max}$/applied stress $\square_0$) shoots upwards drastically from the familiar value of SCF=3 for an isotropic plate. G for some cross-ply carbon fiber composites has been reported as low as 0.2 $E/(2(1+\square))$ [6], resulting in stress concentration factors of an order of magnitude around circular defects and holes. Shear modulus degradation due to matrix cracking can exacerbate the problem [2].

Laminate composites like that pictured in Figure 1b are constructed by stacking together laminae (Figure 1a) at angles $!_i$ chosen to optimize a desired property of the resulting laminate composite structure. Schemes for optimizing the lay-up angles $!_i$ are well established in the literature for uniform stress states (e.g. [5])

Stress concentrations can be reduced or enhanced by tailoring material properties [1], but tailoring to reduce stress concentrations can come at the cost of reducing overall strength, since the strength of each lamina varies with direction of stressing. The objective of this work is an understanding of how laminate optimization changes in the presence of a circular hole. An evolutionary scheme was developed to this end, and optimization was illustrated on a model material system.

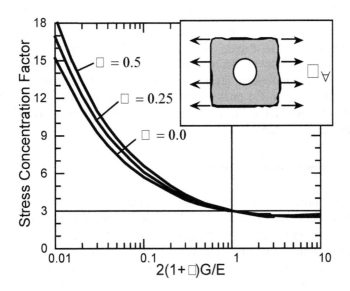

Figure 2. The stress concentration at the boundary of a hole in an infinite plate can be amplified or reduced by anisotropy. Here, the shear modulus is varied relative to the Young's moduli for the directions parallel and perpendicular to the loading direction.

2.   PROBLEM FORMULATION

2.1   Laminate design

In a set of xy axes aligned with the fibers as in Figure 1a, the in-plane mechanical behavior of an orthotropic lamina is characterized by four elastic constants (e.g. [7]):

$$\begin{bmatrix} \Box_{xx} \\ \Box_{yy} \\ \Box_{xy} \end{bmatrix} = \begin{bmatrix} Q_{11} & Q_{12} & 0 \\ Q_{12} & Q_{22} & 0 \\ 0 & 0 & Q_{66} \end{bmatrix} \begin{bmatrix} \Box_{xx} \\ \Box_{yy} \\ \Box_{xy} \end{bmatrix} \tag{1}$$

where $\square_{ij}$ and $\square_j$ are stress and strain components, and Q_{ij} is the stiffness matrix of the lamina.

The in-plane stiffness matrix $\overline{Q}_{ij}$ of a general laminate, consisting of several bonded orthotropic laminae in prescribed orientations (the "local" 1-2 axes defined by l in Figure 1b), can be fully anisotropic. The 6 independent components of the matrix are found through a thickness-weighted sum of the individual layer stiffness matrices Q_{ij}, transformed into the "global" xy axes. In the presence of a stress-concentrating feature like the circular hole pictured in Figure 2, the stress distribution within a linear elastic laminate is dictated by $\overline{Q}_{ij}$.

2.2   Design Optimization

Design optimization involves choosing lamina orientations l_i that maximize strength in the presence of such features. Optimization proceeded with a genetic algorithm (GA); each genome represented the orientations l_i of the laminae within a possible laminate. For every genome in the pool, stresses in each lamina were calculated near the stress concentration, a failure criterion was applied, and a raw score equal to the maximum sustainable force was assigned. Once raw scores were calculated for every genome in the population, an evolution scheme was applied in which new members were added to the population pool by reproduction, mutation and crossover; traits of the good genomes passed to the next generation, while those of the bad ones did not. The population's average fitness was improved until a termination criterion was met.

3.   MODEL PROBLEM

3.1   Problem Description

The case of a symmetrically stacked laminate plate containing a circular hole was considered as an illustration. The hole was large compared to the fiber size but small compared to the dimensions of the plate. For this case, the maximum stress occurred on the hole boundary and the stress distribution can be written in closed form. Lekhnitskii [8] writes the distribution of the thickness-averaged tangential stress along the hole boundary as:

$$\frac{\sigma_{\theta\theta}}{p} = \sin^2\theta + \text{Re}\left\{\frac{e^{i\theta}\left[\cos^3\theta + (\alpha_1+\alpha_2)\sin^3\theta + \left(2-\alpha_1\alpha_2\sin^2\theta\cos\theta\right)\right]}{\left(\sin\theta - \alpha_1\cos\theta\right)\left(\sin\theta - \alpha_2\cos\theta\right)}\right\},$$

$$(2)$$

in which α_1 and α_2 are the two roots with positive imaginary parts from the polynomial:

$$q_1\alpha^4 - 2c_{13}\alpha^3 + (2c_{12}+c_{33})\alpha^2 - 2c_{23}\alpha + c_{22} = 0, \qquad (3)$$

where c_{ij} is the inverse of $\overline{Q}_{ij}$. $\theta = 0$ corresponds to the loading direction in Figure 2.

Strains were calculated from the average stresses by assuming the laminate to be so thin that all laminae deformed identically, which, while not true on the hole boundary itself, is true very close to the hole boundary (e.g. [3])

The strength and stiffness properties used were those of a graphite/epoxy composite currently in use in aircraft manufactured by The Boeing Company [10]. Relative to the direction perpendicular to the fibers, these laminae are strong and stiff in the fiber direction; the shear failure strain is high compared to the normal failure strains. Referring to the axes in Figure 1a, the moduli ratios used were $E_y/E_x = 0.105$ and $G/E_x = 0.0468$; $\nu_{xy} = 0.3$. The failure envelope employed was a maximum strain-type condition, in which failure was said to occur when any of the following criteria were met at any point on the hole boundary:

☐ $\varepsilon_{xx} \geq \varepsilon_0$ *in tension*

☐ $\varepsilon_{xx} \leq -1.15\,\varepsilon_0$ *in compression*

☐ $|\varepsilon_{yy}| \geq 0.368\,\varepsilon_0$

☐ $|\varepsilon_{xy}| \geq 2.04\,\varepsilon_0$

The actual magnitude of ε_0 is unimportant for the optimization.

The procedure of Section 2.2 was employed to find the configuration in which a laminate comprised of a particular number of identical layers has an optimal strength per unit weight.

3.2   Results

The optimal design of symmetric laminates of a range of thicknesses (measured in number of laminae) was studied. The trade-off was the following. Reducing a laminate's anisotropy by turning a lamina away from the direction of loading has the advantage of reducing the strain concentration at the hole boundary; however, this comes at the cost of reducing strength in the rotated lamina at the point of highest stress, typically near the "equator" of the hole.

For the simplest symmetric laminate (two laminae), the optimal material is formed by aligning both laminae with the loading direction. The next simplest symmetric laminate is that constructed with two pairs of laminae. The optimization landscape for this case is illustrated in Figure 3, which shows the failure load for all possible orientations of the inner and outer pairs of laminae. The optimum is found on the 'rim' of the resulting 'volcano.' In this case, tailoring lamina orientations increases the laminate strength only slightly. Note that the landscape even for this simple case contains several local maxima.

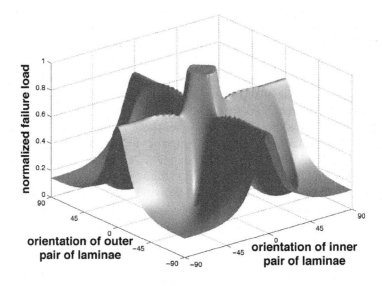

Figure 3. Failure load as a function of lamina orientations for a four-ply symmetric laminate.

As the number of laminae in a laminate increases, the advantage derived from tailoring increases. As shown in Figure 4, the strength per unit weight of the laminate increases as more layers are added in the optimal orientations. For a typical 20-ply laminate, 25% strength improvement is derived from tailoring. Beyond a critical thickness, the optimal laminate has one pair of laminae turned perpendicular to the loading direction, and all others parallel to the loading direction; one pair of laminae is sacrificed to maximally reduce the strain concentration. The dashed line in Figure 4 corresponding to a laminate with a second pair of laminae rotated 90° from the loading direction crosses the line corresponding to a single rotated pair of plies somewhere beyond the right end of the plot; this indicates that the optimal lay-up for a sufficiently thick composite will differ from $[0_{n-2}/90_2]$. However, for all reasonable thicknesses of laminates with this lay-up, the optimum tolerance to circular flaws is reached by turning one pair of laminae 90° to the loading direction.

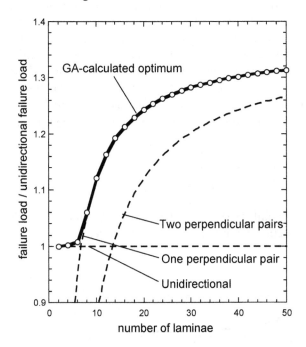

Figure 4 Normalized strength to weight ratio for optimally designed symmetric laminates. Over a broad range of thicknesses, the optimum strength in the presence of a hole is achieved by rotating one pair of laminae 90° from the direction of loading.

4. DISCUSSION

A scheme for optimally tailoring a linear elastic laminate plate for maximum flaw tolerance was developed. Genetic algorithms are particularly well-suited to this scheme, because of the number of local maxima which need to be navigated. The example problem worked showed what in retrospect was a result that could have been predicted, that for laminae strong and stiff in one direction, the optimum configuration over a broad range of thicknesses is reached by rotating a few plies perpendicular to the loading direction.

Is this worthwhile when designing a composite structure? Flaw tolerance is vital to the design of composite structures whose microstructures are inherently random (e.g. [9]) For the case when a small, circular hole is a design necessity in a plate in tension, the procedure followed in this paper yields the design that should obviously be followed. For the case when a circular hole is a defect that might occur between structural inspections during a plate's service lifetime, the decision to tailor a laminate for optimal flaw tolerance involves weighing the benefit derived from tailoring against the loss of strength:weight ratio in the absence of a hole. A critical strain failure criterion predicts that turning one pair of laminae 90° while leaving N-2 laminae aligned in the loading direction reduces a laminate's tensile strength $\Box_{max}$ relative to the strength of a laminate with all layers aligned, $\Box_{max}^{aligned}$. We derive the following expression for this reduced strength:

$$\frac{\Box_{max}}{\Box_{max}^{aligned}} = \left[1 \Box \frac{2}{N}\left(1 \Box \frac{E_y}{E_x}\right)\right]\frac{\Box_y^{crit}}{\Box_x^{crit}}. \tag{5}$$

The critical tensile failure strains and values of E_x and E_y for the model system listed in Section 3.1 are inserted into this equation in Figure 5.

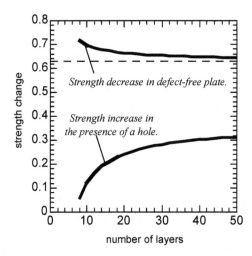

Figure 5. Design for flaw tolerance can increase strength in the presence of a flaw, while decreasing strength in cases when no flaw exists.

As can be seen from Figure 5, Equation (5) predicts a drastic drop in the strength of a plate that has been designed to maintain strength in the presence of a circular defect. In a uniform plate containing no hole, the pair of laminae that is rotated 90° is always the first to crack. This occurs at 37 percent of the failure strain of the laminae that are aligned with the load, meaning that 63 percent is the minimum possible strength reduction in the aligned plate according to the failure criterion. Even after these 90° laminae crack, however, they continue to affect the stress distribution around defects in the laminate. The effect of the residual, post-cracking notch-sensitivity of such a laminate needs further exploration.

ACKNOWLEDGMENTS

Z. Hu and M. J. Jakiela thank the Boeing Foundation for financial support. M. J. Jakiela also gratefully acknowledges the support he receives as the Hunter Professor of Mechanical Design. G. M. Genin thanks the Johanna D. Bemis Trust and the School of Engineering and Applied Science at Washington University in St. Louis.

References

[1] Budiansky, B., J.W. Hutchinson, and A.G. Evans, "On Neutral Holes in Tailored, Layered Sheets," *J. Applied Mechanics*, **60**, 1993, 1056-7.

[2] Genin, G.M., and J.W. Hutchinson, "Composite Laminates in Plane Stress: Constitutive Modeling and Stress Redistribution Due to Matrix Cracking," *Journal of the American Ceramic Society*, **80**(5), 1997, p. 1245-55.

[3] Genin, G.M., and J.W. Hutchinson, "Failures at Attachment Holes in Brittle Matrix Laminates," *Journal of Composite Materials*, **33**(17), 1999, 1600-1619.

[4] Green, A.E., and W. Zerna, *Theoretical Elasticity*, London: Oxford University Press, 1968, 350-351.

[5] Gürdal, Z., R.T. Haftka, and P. Hajela, *Design and Optimization of Laminated Composite Materials*, New York: Wiley, 1999.

[6] Heredia, F.E., S.M. Spearing, T.J. Mackin, M.Y. He, A.G. Evans, P. Mosher, and P. Brønsted, "Notch Effects in Carbon Matrix Composites," *Journal of the American Ceramic Society*, **77**, 2817-2827 (1994).

[7] Jones, R., *Mechanics of Composite Materials*, New York: McGraw-Hill, 1975.
Lekhnitskii, S.G., *Anisotropic Plates*, Trans. by S. W. Tsai and T. Cheron, New York: Gordon and Breach, 1968.

[8] McNulty, J., F. W. Zok, G. M. Genin, and A. G. Evans, "Notch-Sensitivity of Fiber-Reinforced Ceramic Matrix Composites: Effects of Inelastic Straining and Volume-Dependent Strength," *Journal of the American Ceramic Society*, **82**(5), 1999, 1217-28.

[9] Rubin, A.M., unpublished material data, The Boeing Company, 2000.

[10] Savin, G.N., *Stress Concentration around Holes*, Trans. by W. Johnson. New York: Pergamon, 1961.

GENETIC ALGORITHM FOR DAMAGE ASSESSMENT

V.T. Johnson[1], S. Anantha Ramu (Late)[2] and B.K. Raghu Prasad[3]
[1]*Infosys Technologies Limited, Bangalore.*
[2]*Professor (Late), Department of Civil Engineering, Indian Institute of Science, Bangalore.*
[3]*Professor, Department of Civil Engineering, Indian Institute of Science, Bangalore.*
bkr@civil.iisc.ernet.in

Abstract: Genetic Algorithms (GAs) are recognized as an alternative class of
            computational model, which mimic natural evolution to solve problems in a
            wide domain including machine learning, music generation, genetic synthesis
            etc. In the present study Genetic Algorithm has been employed to obtain
            damage assessment of composite structural elements. It is considered that a
            state of damage can be modeled as reduction in stiffness. The task is to
            determine the magnitude and location of damage. In a composite plate that is
            discretized into a set of finite elements, if a jth element is damaged, the GA
            based technique will predict the reduction in E_x and E_y and the location j. The
            fact that the natural frequency decreases with decrease in stiffness is made use
            of in the method. The natural frequency of any two modes of the damaged
            plates for the assumed damage parameters is facilitated by the use of Eigen
            sensitivity analysis. The Eigen value sensitivities are the derivatives of the
            Eigen values with respect to certain design parameters. If $\square_{iu}$ is the natural
            frequency of the ith mode of the undamaged plate and $\square_{id}$ is that of the
            damaged plate, with $\square_i$ as the difference between the two, while $\square\square_k$ is a
            similar difference in the kth mode, R is defined as the   ratio of the two. For a
            random selection of E_x, E_y and j, a ratio R_i is obtained. A proper combination
            of E_x, E_y and j which makes $R_i - R = 0$ is obtained by Genetic Algorithm.

Keywords: genetic algorithm, damage assessment, NDE of composites, vibration
            technique.

1.   INTRODUCTION

Genetic Algorithm search methods are rooted in the mechanism of
evaluation and natural genetics. The basic idea is to maintain a population of

131

T. Burczyński and A. Osyczka (eds),
IUTAM Symposium on Evolutionary Methods in Mechanics, 131–141.
© *2004 Kluwer Academic Publishers.*

knowledge structures that represents candidate solutions to the current problem. Each member of the population will ideally be a binary string of fixed length in which the problem specific information is encoded. The population evolves over time through competition (survival of the fittest) and controlled variation (recombination and mutation). Theoretical analysis have shown that GA exploits the knowledge accumulated during search in a way that efficiently balances the need to explore new areas of the search space with the need to focus on high performance regions of the search space.

The basic paradigm is shown in Figure 1 which is based on the method proposed by Holland[3] and later extended by Goldberg[4].

The population can be initialized using whatever knowledge is available about the possible solutions. In the absence of such knowledge, the initial population should represent a random sample of the search space.

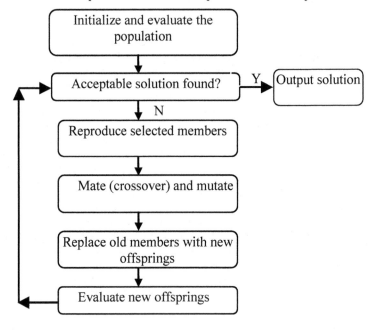

Figure 1. Paradigm of GA

Each member of the population is evaluated and assigned a measure of its fitness as a solution. When each member in the population has been evaluated, a new population is formed using genetic operators. The usual genetic operators are : selection, crossover, mutation, dominance, inversion, intra chromosomal duplication, deletion, translocation etc. Many different algorithms have been proposed which utilize several of these operators. The simplest among these known as Simple Genetic Algorithm (SGA), uses selection, crossover, and mutation as the genetic operators, and the same is proposed for the present study.

2.   PROBLEM FORMULATION

It is considered that a state of damage can be modeled as a *reduction in stiffness* or in other words as reduction in Young's modulus, which may be local (or global). Now, the task is to determine the severity of the damage and its location.

A laminated composite plate is considered. The plate is divided into a number of small elements. Let only the jth element be damaged and let E_x* and E_y* be the damaged moduli for that element. The problem can be stated as : Can a GA based system predict the reduction in E_x , E_y and the location j at which this damage has occurred?

The fact that the ratio of the change in natural frequency (between undamaged and damaged) for any two distinct modes is only a function of the position of the element damaged (i.e., local damage), is made use of in formulating this problem. This can be shown as follows[1].

The change in the natural frequency $\Box_i$ of mode i of a structure due to localized damage is a function of the position vector of the damage, *r,* and the reduction in stiffness caused by the damage, $\Box K$. Thus

$$\Box_i = f(\Box K, \mathbf{r}) \tag{1}$$

Expanding this function about the undamaged state ($\Box K$=0), and ignoring second-order terms, yields

$$\Box_i = f(0, r) + \Box K \frac{\Box f}{\Box(\Box K)}(0, r) \tag{2}$$

But f(0,**r**) = 0, for all **r**, since there is no frequency change without damage. Hence,

$$\Delta\omega_i = \Delta K\, g_i(r) \tag{3}$$

Similarly,

$$\Delta\omega_j = \Delta K\, g_j(r) \tag{4}$$

Thus, provided that the change in stiffness is independent of frequency,

$$\frac{\Delta\omega_i}{\Delta\omega_j} = \frac{g_i(r)}{g_j(r)} = h(r) \tag{5}$$

That is, the ratio of the frequency changes in two modes is therefore only a function of the damage   location. Positions where the theoretically determined   ratio ($\Delta\omega_i / \Delta\omega_j$) equals the experimentally measured value are therefore possible damage sites. Suppose the natural frequency of the undamaged plate is known. Then, knowing the natural frequency of the damaged plate, the ratio of change in natural frequency of the damaged plate for two modes can be calculated. Now, the above problem can be restated as:

Given the ratio of change in natural frequency for two modes of the damaged plate, can we find out the position **j** of the damaged element and the value of E_x and E_y in the damaged state, assuming all other elements are undamaged?

That is, now the problem addressed is a *search* problem - a search for the right combination of E_x, E_y and **j** such that the new system satisfies the given ratio. Mathematically, this can be formulated as follows:

Let ω_{iu} be the natural frequency of ith mode of the undamaged plate and let ω_{id} be the natural frequency of ith mode of the damaged plate under consideration. Define

$$\Delta_t = \frac{\Delta\omega_i}{\Delta\omega_k} \qquad \text{where } \Delta\omega_i = \omega_{iu} - \omega_{id} \tag{6}$$

For a random selection of E_x, E_y and **j**, the ratio defined as above can be evaluated for the given plate. Let this be Δ_c. Now the problem reduces to find out the right combination of E_x, E_y and **j** for which $\Delta_t - \Delta_c = 0$ (equivalently $\Delta_t / \Delta_c =1$). A solution to a similar search problem is given by Cawley and Adams[1] which involves dynamic FEM and a slightly involved procedure but at the end it can only give the location not the degree of damage and Johnson[4] which is based on Artificial Neural Network(ANN) based approach. Here, the above search is proposed to be carried out through GA which can give the location and the degree of damage simultaneously.

3.   PROCEDURE

It is assumed that the natural frequency of the structure in the undamaged state as well as in the damaged state is given. Compare both and if there is a difference follow the procedure given below to locate and quantify the damage.

- Discretize the structure into a finite number of elements (N). Calculate $\Box_t$ as defined in Eqn. 6.

$$\Box_{t1} = \frac{\Box_1}{\Box_2} \quad \text{and} \quad \Box_{t2} = \frac{\Box_3}{\Box_4} \quad \text{where} \quad \Box_i = \Box_{iu} \Box \Box_{id} \qquad (7)$$

and **i** represents the mode number.

- Initialize a population consisting of an adequate number of members, (30 to 70 is recommended), which represents possible solution sets. Each member is represented as a string formed by encoding the reduction factor of E_x ($\Box$) & E_y ($\Box$) and the position. The first two values are real and continuous, which varies from 0.0 to 1.0 whereas the third value is taken as the element number and is hence an integer, which varies from 1 to N. Each of these values can be appropriately encoded as a binary string. A discrete approach is used for coding the $\Box$ and $\Box$ values. The range 0-1 is discretized into a fixed number of levels. That is, if the length of the substring for representing $\Box$ or $\Box$ is fixed as l, there are 2^l possible combinations for which the decimal equivalent value of the substring spans the range $0 - (2^l -1)$ and it is normalized by dividing by its maximum decimal equivalent value possible $(2^l -1)$ so that the resulting value lies in the range 0-1. The three substrings are concatenated to form a population member. If $\Box$ and $\Box$ and position are represented by strings of length l_1, l_2 and l_3 bits respectively, the length of each population member (L) is given by $L = l_1 + l_2 + l_3$.
- Decode each member to get the $\Box$, $\Box$ and j (position) and calculate the $\Box_{c1}$ and $\Box_{c2}$ as defined below.

$$\Box_{c1} = \frac{\Box_1}{\Box_2} \quad \text{and} \quad \Box_{c2} = \frac{\Box_3}{\Box_4} \quad \text{where} \quad \Box_i = \Box_{iu} \Box \Box_{id} \qquad (8)$$

and **i** represents the mode number.

— Calculate the fitness function which is defined as

$$f_1 = \frac{\Box_{c1}}{\Box_{t1}} \quad \text{and} \quad f_2 = \frac{\Box_{c2}}{\Box_{t2}} \tag{9}$$

where $\Box_{t1}$ and $\Box_{t2}$ are the target values supplied by the user (Eqn. 7). The value of f_1 and f_2 are further tuned by the following scaling:

$$f_1 \# \quad f_1 \qquad \text{if } 0< f_1 <1.0 \tag{10}$$

$$f_1 \# \quad 2.0 - f_1 \text{ if } 1.0< f_1 <1.5 \tag{11}$$

$$f_1 \# \quad 0.1 \qquad \text{if } f_1 > 1.5 \tag{12}$$

Scaling is done in the same way for f_2 also. Then the fitness value is defined as

$$F_i = (f_1 + f_2) / 2.0 \tag{13}$$

where F_i denotes the fitness value of the ith member of the population.

— Create a mate pool as follows: sort the population in ascending order of the fitness value. Then calculate

$$n = \frac{F_i}{\bar{f}} \tag{14}$$

for each member, where $\bar{f}$ is the average of fitness values for all members. The number of offsprings to be reproduced for each member is determined by rounding off the above quantity to the nearest integer. Clearly all members having n less than 0.5 will die off. If the total number of offsprings thus produced does not match with the population size (P_{size}), one offspring each is given to members from the first member onwards until the total population size is reached.

— Create a new pool by applying the crossover operator as follows: pick a copy of member #0 (members are counted from 0 onwards) from the mating pool (resulted after reproduction) and select a mate from the pool randomly (i.e., generate a random number between 0 and (P_{size}-1)). These two members are subjected to crossover to produce two children if a random number generated in the range 0 to 1 is less than the crossover rate, P_c (The recommended range for P_c is 0.7- 1.0). Otherwise the strings remain unaltered. Among the different crossover operators such as one-point crossover, two-point crossover, uniform crossover etc., one-point crossover is implemented for the present work. Assuming that l is the string length, it randomly chooses a crossover point that can assume values in the range 0 to l-1 (bit locations are labeled from 0 to l-1). The portions of the two strings beyond this crossover point are exchanged to form two new strings (children). The resulted child strings constitute the first two members (member #0 and #1) in the new pool.

For creating member #2 and #3 of new pool, pick member #2 from the mating pool and select a mate randomly from the mating pool and subject it to crossover. Repeat this for alternate members of the mating pool so that a new pool of members (P_{size} remains constant) is formed.

– Apply mutation operator as follows: each bit of the string is flipped (changing a 0 to 1 or vice versa) independently if a random number generated in the range 0-1 is less than the mutation rate P_m (The recommended value for P_m is 0.001 to 0.1. The SGA treats mutation only as a secondary operator with the role of restoring lost genetic material. For example, suppose all the strings in a population have converged to a 0 at a given position and the optimal solution has a 1 at that position. Then crossover cannot regenerate a 1 at that position, while a mutation can.).

– The new population generated after mutation replaces the previous generation population and the above procedure is repeated until an acceptable solution is found. A solution produced is accepted if both f1 and f2 for a member are greater than 0.99.

– Terminate the process if the number of generations reaches 100 (case where an acceptable solution was not found within a reasonable number of generations). The whole exercise is repeated for another set of parameters(P_{size}, P_c and P_m).

It is to be noted that the evaluation of a member needs the calculation of change in natural frequency for the damage parameters it represents. The calculation of the change in natural frequency for the assumed damage parameters is greatly facilitated with the use of Eigen sensitivity analysis which otherwise needs the costly dynamic FEM computation at every step. This is very essential as each generation contains many possible solutions and if a FEM analysis is to be done for each member in a given generation, the efficiency of GA cannot be realized in any case.

4.   EXAMPLE PROBLEM

The strategy and procedure discussed in the previous section for the GA-based damage assessment problem is illustrated below. A composite plate made of T300-N5208, with geometric dimension 0.5m x 0.5m x 0.002m and with properties as given in Table 1 is selected. The plate is discretized into 100 elements which are numbered as shown in Figure2. The natural frequencies of the plate in the undamaged state is worked out as given in Table 2.

91	92	93	94	95	96	97	98	99	100
81	82	83	84	85	86	87	88	89	90
71	72	73	74	75	76	77	78	79	80
61	62	63	64	65	66	67	68	69	70
51	52	53	54	55	56	57	58	59	60
41	42	43	44	45	46	47	48	49	50
31	32	33	34	35	36	37	38	39	40
21	22	23	24	25	26	27	28	29	30
11	12	13	14	15	16	17	18	19	20
1	2	3	4	5	6	7	8	9	10

Figure 2. Discretization of the plate (0.5m x 0.5m)

For the illustration purpose, the plate is damaged at the element number 26 to the extent 0.4 and 0.6 for E_x and E_y respectively (i.e., E_x and E_y are reduced to 0.4 and 0.6 times their original value at the location 26. See Figure 2). The corresponding natural frequencies of the plate, worked out through the sensitivity analysis is also provided in Table 2 (column 3). $\square_{t1}$ & $\square_{t2}$ are calculated as defined in Eqn. 7 which come to 0.0995 and 0.303 respectively.

Table 1. Plate properties

Property	Value (N/m^2)
E_x	86.9 x 10^6
E_y	48.7 x 10^6
G_{xy}	26.03 x 10^6
V_{xy}	0.411

Table 2. Natural frequencies of the plate

Mode #	□ - undamaged (rad/sec)	□ - damaged (rad/sec)
1	571.1491	570.72
2	1068.596	1064.28
3	1266.632	1265.86
4	1768.168	1765.62

The search variables are the □, □, and position, where □ is the reduction in E_x and □ is the reduction in E_y. A binary string of length 17 (17 bits) is formed (Figure.3).

0	1	2	3	4	5	6	7	8	9	10	11	12	13	14	15	16
0	0	1	1	1	1	0	1	1	1	1	0	1	0	1	0	0

Figure 3. Typical member of the population

First 5 bits represent □, the next 5 bits represent □ and the last 7 bits represent the position of damage (element number). Exact values of □ and □ are calculated by dividing the integer value of the first 5 bits and that of the next 5 bits by 2^5 respectively. The position is obtained by calculating the integer value of the last 7 bits.

5. DISCUSSION

The application of GA, reported herein, shows its potential for system identification problems. The problem was formulated in a simplified way by adopting two assumptions - the damage was defined as reduction in stiffness which is the consequence of different damage modes, and only one element was allowed to be damaged. The exercise is essentially of iterative nature but better and better knowledge is extracted after each generations based on the previous experience. This makes GA distinct from other techniques. The critical factor is having an appropriate evaluation function. Quite often, some kind of simulation is needed to arrive at this, for example, evaluating the natural frequencies of the plate under the assumed damage state. Provided there is a suitable evaluation function, the strategy is independent of type, geometry and other parameters of the structure. This gives another potential point to GA compared to the ANN approach where the strategy is very much dependent on the structure. i.e., to devise a *learnt ANN* for a general class of structure is extremely difficult, if not impossible. But the advantage with ANN approach is the instantaneous nature of the output from a learnt network, which makes it suitable for real time monitoring.

Though the plate considered in the example problem is symmetrical, the symmetry is not considered (The ratios defined will be same for symmetric elements). This is in order to make the solution space complex for illustration purpose. This point also shows that the strategy adopted presently can result in a loci of candidate locations for a symmetric structure.

Experiment with other damage location and damage severity was repeated for 15 cases, results for which are presented in Table 3. It can be seen that, in all the cases, position of the damaged element was located exactly and the estimated damage severity was closer to the actual value. Further attention was paid to the effect of the GA specific parameters - Population size, Crossover rate, and Mutation rate on the performance of the GA based system. Another example is chosen, in which the 88th element is considered to be damaged to the extent $\square = 0.6$ and $\square = 0.4$. The P_{size}, P_c and P_m are varied and the experiment is repeated to study the effect of each of the parameters on the performance of GA. In each case, two plots are generated: generations vs. best fitness, and generations vs. improvement. Best fitness is the largest fitness value attained within a generation. Improvement is defined as the ratio of the average fitness of the current population to the average fitness of the previous population. Average fitness is the average of the fitness of all members within a generation.

Table 3. Typical results

Sl.No	Actual values (el. no, $\Box$, $\Box$)	Obtained values (el. no, $\Box$, $\Box$)	No. of generations	Remarks
1	(37, 0.55, 0.48)	(37,0.548, 0.484)	4	$P_{size} = 30$
2	(46, 0.6, 0.8)	(46, 0.677, 0.838)	14	$P_c = 0.8$
3	(75, 0.55, 0.75)	(75, 0.483, 0.64)	53	$P_m = 0.05$
4	(78, 0.71, 0.81)	(78, 0.87, 0.94)	62	for all cases
5	(56,0.75,0.55)	(56,0.678,0.355)	14	
6	(6, 0.88, 0.44)	(6,0.839,0.419)	11	
7	(70, 0.66, 0.33)	(70, 0.87, 0.484)	5	
8	(70, 0.57, 0.87)	(70, 0.6, 0.52)	16	
9	(97, 0.82, 0.72)	(97,0.9,0.65)	34	
10	(88,0.63,0.35)	(88,0.61,0.45)	12	
11	(73, 0.8, 0.8)	(73, 0.678, 0.48)	32	
12	(33, 0.4, 0.7)	(33, 0.29, 0.6)	23	
13	(52, 0.55, 0.75)	(52, 0.163,0.452)	8	
14	(78, 0.6, 0.7)	(78, 0.645,0.742)	40	
15	(78, 0.71, 0.81)	(78, 0.87, 0.94)	62	

6.   CONCLUSIONS

The problem that is solved with GA is complex. Based on the change in measurable system properties (natural frequencies here), it is required to locate the damage and assess the extent of damage. With the novel formulation of this problem, capability of GA has been demonstrated to carry out the above task very efficiently which affirms the possibility of a wider application domain for GA.

References

[1] Cawley,P. and Adams,R.D., "The location of defects in structures from measurement of natural frequencies", *J. Strain Analysis*, Vol.14, No.2,1979, pp.49-57.
[2] Goldberg,D.E., *Genetic algorithms in search, optimization and machine learning*, Addison-Wesley, Reading, Mass., 1989.
[3] Holland,J., *Adaptation in natural and artificial systems*, Univ. Michigan Press, Ann Arbor, Michigan, 1975.
[4] Johnson,V.T., *Biologically Inspired Computational Methods(BICMs) in the structural design and damage Assessment of composites*, PhD thesis, Indian Institute of Science, 1997.

ESTIMATION OF PARAMETERS FOR A HYDRODYNAMIC TRANSMISSION SYSTEM MATHEMATICAL MODEL WITH THE APPLICATION OF GENETIC ALGORITHM

Andrzej Kęsy, Zbigniew Kęsy, Arkadiusz Kądziela
Institute of Applied Mechanics, Technical University of Radom, Poland.
E-mail: akesy@pr.radom.pl

Abstract: In this paper application of genetic algorithm for estimation of parameters for a mathematical model of hydrodynamic transmission system was introduced. The measurements needed for estimation procedure were performed on the test rig. The modeling errors of hydrodynamic transmission system obtained by using genetic algorithm are comparable with the values which were obtained by using the Monte Carlo method. As a result it was concluded that a genetic algorithm can be successfully applied to identify hydrodynamic transmission system.

Keywords: Genetic algorithm, identification, estimation of parameters, hydrodynamic transmission system.

1. INTRODUCTION

Hydrodynamic transmission system (HTS) is a system utilizing hydrodynamic torque converter (HTC) as the means of matching characteristics of the engine to characteristics of machinery and vehicle traction. Due to HTC's ability to automatically adjust to the applied load this type of transmission system is particularly useful in devices that work under constantly changing load and operating conditions. Additionally HTC decreases dynamic loads and dampens torsional vibrations which significantly improves the durability of the device. The design problems of new drive systems of machinery and vehicles including new HTS are still open issues mostly because of introducing to the market ecological engines: combustion engines with low rotational speeds or engines fueled by

143

T. Burczyński and A. Osyczka (eds),
IUTAM Symposium on Evolutionary Methods in Mechanics, 143–152.
© 2004 *Kluwer Academic Publishers.*

alternative fuels such as alcohol or colza oil. Characteristics of these engines are significantly different than characteristics of currently used engines. Therefore it forces development of new constructions of HTS. Identification methods are used during a design process of HTS.

Identification process of mechanical objects contains modeling, experiment, estimation and verification.

Industrial experience shows that practically there is a significant discrepancy between actual characteristics of HTS and characteristics assumed by the designer. The degree of discrepancy depends on obtained precision of applied mathematical model and on errors created during identification of mathematical model which are caused by large fabrication errors of bladed wheels.

One of the effective methods for improving the precision of mathematical modeling is careful estimation of model parameters.

This paper compares two estimation methods of mathematical model of HTS, Monte Carlo method and genetic algorithm method. An analysis was performed for two HTS modeled on test rig equipped with HTCs, which differed in size and type of working fluid flow through the bladed wheels.

2.   MATHEMATICAL MODEL OF HTC

Assuming that shaft flexibility is negligible, a steady state motion of HTS can be described by two torque equations such as:

$$M_s = M_1, \quad M_r = M_2, \tag{1}$$

where: M_s - torque of driving engine, M_r - torque of external resistance, M_1, M_2 - bladed wheels motion torque of HTC's pump and turbine.

Torques of bladed wheels are developed based on theory of average stream under typical assumptions [1, 2]. It is assumed that fluid flow in working space of HTC is a single-dimensional stream, that flows along average line of bladed wheel channels and has average parameter values of all streams which in turns give overall fluid flow. Furthermore in order to increase precision the following were assumed [3]:

- ☐ Viscosity and density of working fluid depend on fluid temperature.
- ☐ Volume of stream of working fluid is constant.
- ☐ Inlet and outlet angles of fluid flow from channels of bladed wheels are in agreement with the curve shape of blades.
- ☐ Fluid flow losses in channels determined for i-nth bladed wheel by a coefficient of flow losses ☐ are dependent on the type of flow, whether the flow is laminar or turbulent.

□   Stroke losses of fluid stream against blades, determined for i-nth bladed wheel by coefficient of stroke losses $!_i$ are dependent linearly on stroke angle $\square_i$ according to equation:

$$!_i = a_{u,i} + b_{u,i}\square_{u,i},\qquad (2)$$

where: $a_{u,i}, b_{u,i}$ – constant coefficients.

An equation of power balance of HTC is usually added to the torque equations. After considering the components of motion torque of bladed wheels, the mathematical model of HTS is taking a form of [2]:

$$
\begin{aligned}
M_s &= M_{h1} + M_{d1} + M_{t1}\\
M_r &= M_{h2}\,\square\,M_{d2}\,\square\,M_{t2}\\
&\square\left(a_1\square_1^2 + a_2\square_2^2 + a_3Q^2 + a_4\square_1Q + a_5\square_2Q + a_6\square_1\square_2\right)\square P_{ch}/Q = 0
\end{aligned}\qquad (3)
$$

where: M_{h1}, M_{h1} – hydraulic torques created by fluid flow in working space of bladed wheels, M_{d1}, M_{d2} - torques of rotating elements submerged in fluid, M_{t1}, M_{t2}, - torques of friction in bearings, ω_1, ω_2 - angular velocities of bladed wheels of turbine and pump, Q - volumetric fluid current in working area, ρ - density of working fluid, a_1,....., a_6 - constants dependent on working area of bladed wheels, P_{ch} - flow losses in channels of bladed wheels.

In the model, three of the above mentioned elements can be taken as independent variables: angular velocity of bladed wheels of pump and turbine ω_1, ω_2,, torques M_s, M_r and volumetric fluid current flow in working area Q.

3.   EXPERIMENTAL STUDY OF HTC

In order to estimate parameters of mathematical model of HTS experimental studies were performed for two HTS systems with HTC type PH-280 and U 358011E. In this study typical experimental stations modeling steady state motion of HTS were used. Applied HTCs differed in active diameters, flow through bladed wheel of turbine and curve run (increasing and decreasing) of torque coefficient $\lambda = f(i_k)$ determined by equation:

$$\lambda = M_s / \rho D^5 \omega_1^2 \qquad (4)$$

where: $i_k = \omega_2 / \omega_1$ - speed ratio of HTC, D - active diameter of HTC.
The above mentioned differences resulted with different modeling accuracies of HTS with these HTCs.

In this experiment rotational velocities of input (ω_1) and output (ω_2) shaft of TS and average inlet temperature of working fluid to the HTC were assumed as input data. Studies were performed for constant angular velocity of inlet shaft ω_1 equal to nominal velocity of HTS for three temperature ranges of working fluid. For each temperature range full cycle study was performed. During the study cycle the angular velocity of input shaft remained constant and the angular velocity of output shaft was changing in such a way that the already assumed values of speed ratio i_k were obtained. For each measuring point predetermined in the above way, torques on input and output shaft and HTC inlet temperature T of working fluid were recorded. Measurements were performed for i_k changes in a range of 0 to 1.

As a result of the experiment steady-state characteristics of HTS were obtained (calculated based on measured data) for assumed temperature range of working fluid in a form of:

$$i_d = f(i_k); \quad \exists = f(i_k); \quad \square = f(i_k) \qquad (5)$$

where: $i_d = M_r / M_s$ - torque ratio, $\exists = i_k i_d$ - efficiency.

4. PARAMETER ESTIMATION OF HTC MODEL

Parameter estimation of a mathematical model of a transmission system depends on obtaining values of selected mathematical model parameters for the predetermined model structure based on the results of the performed experimental study. Estimation was performed based on an algorithm that allows obtain optimum quality of the model determined by model quality criteria.

Up to now, stroke loss coefficient and flow loss coefficient were assumed as the estimated parameters in typical mathematical models of HTS [4]. For the analyzed model the following estimated parameters were assumed [3]:

- □ Flow loss coefficient ε_i, the same for all bladed wheels, but this coefficient takes different values for laminar flow (Re<2300) and turbulent flow (2300≤Re<80,000).
- □ Coefficients $a_{u,i}$, $b_{u,i}$ of linear equation determining stroke loss coefficient Φ_i, but parameter $b_{u,i}$ may have two different values dependent on sign of stroke angle $\alpha_{u,i}$.

☐ Mechanical efficiency of bladed wheels of pump and turbine - $\eta_{m,1}$, $\eta_{m,2}$.

Overall there were 13 parameters.

Values of the estimated parameters remained in the same ranges as parameters used in actual constructions or recommended by literature. For the need of parameter estimation of HTS with HTC type PH-280 model all 13 parameters were used because this HTS is difficult to model. For HTS with HTC type U 358011E the number of parameters were decreased to 7 assuming that coefficients b_{ui} are equal to zero.

The difference between tangent characteristics of TS calculated based on the model and the characteristics obtained based on experimental study was assumed as the quality criteria of the model, defined as:

$$
K = \frac{1}{4}\{\sum_{r=1}^{r_o} [\sum_{m=1}^{m_o} \frac{1}{N_m} \sqrt{\frac{1}{k_o} \sum_{k=1}^{k_o}(y_m^{r,k} \Box \overline{y}_m^{r,k})^2}
$$

$$
+ \sum_{m=1}^{m_o} \frac{1}{N_m} \max_{k=1,k_o} \left| y_m^{r,k} \Box \overline{y}_m^{r,k} \right|]\}
\tag{6}
$$

where: y - output value obtained from the model, $\overline{y}$ - output value obtained from the experiment, m_o - number of output values, r_o - number of studied input data, k_o - number of measurements of output data, N_m - value normalizing m-th output signal. Study case parameter were: $y_1 = M_s$, $y_2 = M_r$, $m_o = 2$; for $m_o = 1$, $N_m = i_d$, for $m_o = 2$ $N_m = 1$. Also for HTS with HTC type PH-280, $k_o = 22$, $r_o = 6$, and for HTC type U 358011E $k_o = 11$, $r_o = 3$. The quality criteria (6) should accept a minimum value.

5. ESTIMATION COURSE AND RESULTS

Static equation of a mathematical model of HTS is complex. Therefore it is impossible to create a simple mathematical formula to define quality criteria as a function of estimated parameters. A solution can be obtained only through application of numerical methods. Monte Carlo method and genetic algorithm were applied for minimization of quality criteria. Calculations were performed based on computer programs written in Turbo Pascal 7.0 programming language.

Monte Carlo method. Application of this method for parameter estimation relied on random selection of parameters from their assumed value ranges. The value of quality criteria was calculated for each set of the selected parameter. From the set of selected parameter values and the values of quality criteria the minimum quality criteria values were recorded and saved. Calculations were aborted when after 1000 consecutive random selections quality criteria did not change. A set of estimated parameters that assured the best approximation of steady state characteristic of studied HTS by mathematical model was considered as the set which gave the smallest value of quality criteria K_{min} from all computer model simulations.

Genetic Algorithm. In this case a genetic algorithm was applied to obtain parameter values assuring minimum value of quality criteria according to the formula from [6].

Values of the estimated parameters were rescaled to a range of 0 to 1. Each parameter value was coded into a ten bit binary word, therefore for each parameter ten bits were assigned within a single binary chromosome. The binary chromosome contained 70 bits for HTS with HTC type U 358011E and 130 bits for HTS with HTC type PH-280. Fitness function F defined as inverted value of quality criteria of our model (6), $F = 1/K$. The tournament method was used for chromosome selection. Crossover probability p_k was assumed as 0.8 or 0.9 and mutation probability p_m was assumed to be 0.01. Calculations were performed on populations containing $N = 50$ and $N = 100$ binary chromosomes. The genetic algorithm computer program was repeatedly run 10 times for each combination of parameters. Calculations were aborted when the fitness function did not change its value for 1000 consecutive iterations. For each studied HTS, the binary chromosome having the highest value of fitness function obtained during all program runs F_{max} (minimum value of quality criteria) was de-coded obtaining optimal set of parameters guaranteeing the best approximation of static characteristic of studied HTS by mathematical model.

Maximum values of the fitness function F_{max} obtained during calculations and generation numbers where they occurred $n_{i,max}$ are provided in Table 1.

Table 1. Maximum values of fitness function obtained during calculations

TS with HTC type	N	p_k	F_{max}	$n_{i,max}$
PH-280	100	0,9	**0,05891520**	1362
PH-280	100	0,8	0,05598753	1400
PH-280	50	0,9	0,05147303	1087
PH –280	50	0,8	0,05081254	1125
U 358011E	100	0,9	**0,17160284**	2257
U 358011E	100	0,8	0,17120800	2202
U 358011E	50	0,9	0,17091729	2186
U 358011E	50	0,8	0,17020988	2107

Figure 1 shows convergence of applied minimization methods of quality criteria. The characteristics graphs were prepared for the computer simulation where minimal quality criteria was obtained.

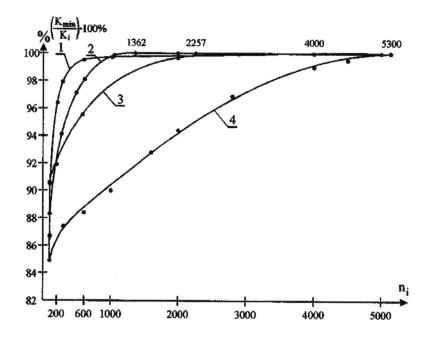

Figure 1. Convergence of minimization methods of quality criteria K for the studied HTS
        1 - Genetic algorithm, HTS with HTC type U 358011E;
        2 - Genetic algorithm, HTS with HTC type PH-280;
        3 - Monte Carlo method, HTS with HTC type PH-280;
        4 - Monte Carlo method, HTS with HTC type U 358011E.

6. VERIFICATION OF CALCULATION RESULTS

In order to evaluate the usefulness of estimation procedure of mathematical model of HTS, verification was performed through comparison of steady-state characteristics curves (5) calculated from mathematical model and the curves obtained from experimental studies. Maximum relative error was assumed as a measure of divergence between the results of the experiment and the results of the model calculations assumed. The maximum relative error was defined as:

$$\square_{x,max} = \max \left| \frac{x^e \square x^m}{x^e} \right| 100\,\%; \qquad (7)$$

where: x - corresponds to $i_d = f(i_k)$ and $\square = f(i_k)$, but superscript index means: m - value obtained from static characteristic calculated based on a model with optimal parameters obtained from estimation for the analyzed i_k, e - value obtained from static characteristic based on experiment for the same i_k.

Table 2 shows maximum relative error values of steady-state characteristics (5) of the studied HTS calculated according to (7) by the application of the Monte Carlo method to minimize the quality criteria.

Table 2. Relative error values $\square_{max}$ of steady-state characteristic of HTS with estimated parameters by Monte Carlo method

$\square_1$ [rad/s]	T range [°C]	$\square_{i_d,max}$ [%]	$\square_{\square,max}$ [%]
HTS with HTC type PH-280			
210	16 – 21	4,42	6,57
	62 – 68	3,57	5,67
	94 – 104	3,68	5,50
Arithmetic average		4,90	
HTS with HTC type U 358011E			
105	20 –21	2,42	5,91
	50 – 52	2,22	3,36
	70 – 75	2,16	3,28
Arithmetic average		3,22	

Table 3 shows the error values obtained using genetic algorithm.

Table 3. Error values $\square_{max}$ of steady-state characteristics of HTS with estimated parameters by genetic algorithm

$\square_1$ [rad/s]	T range [°C]	$\square_{i_d,max}$ [%]	$\square_{\square,max}$ [%]
HTS with HTC type PH-280			
210	16 – 21	4,14	6,78
	62 – 68	4,66	4,81
	94 – 104	3,92	4,65
Arithmetic average		4,83	
HTS with HTC type U 358011E			
105	20 –21	4,73	6,20
	50 – 52	3,16	3,19
	70 – 75	2,89	2,62
Arithmetic average		3,80	

Figure 2 and Figure 3 present examples of steady-state characteristics (5) of HTS with HTC type U 358011E. The theoretical characteristics were calculated by applying in mathematical model optimal parameters obtained from the Monte Carlo method and from the genetic algorithm. The experimental characteristics were calculated based on measured data.

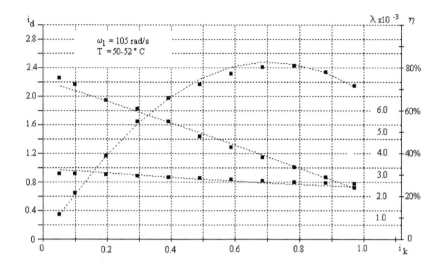

Figure 2. Steady-state characteristic of HTS with HTC type U 358011E:
– – – from mathematical model estimation by applying the Monte Carlo method ,
■ – from experimental studies.

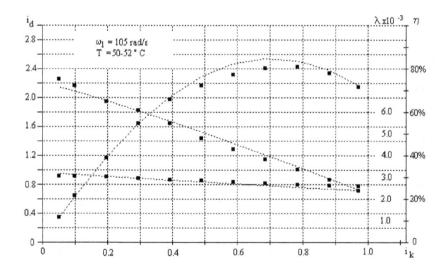

Figure 3 . Steady-state characteristics of HTS with HTC type U 358011E:
– – – from mathematical model estimation by applying the genetic algorithm,
■ – from experimental studies.

7. CONCLUSIONS

Mathematical model of HTS is based on an average stream model which in its basic form doesn't take into account many factors that have significant influence on HTS characteristics. In order to improve modelling accuracy the model was expanded by considering a variety of factors e.g. gap widths between bladed wheels. This type of modification usually doesn't provide expected improvement of modelling accuracy. For example, analysis described in [4] shows maximum modelling error of steady-state characteristics of HTS with HTC type ZM-130 equal to 40%. The error values obtained from HTS modelling with an application of the Monte Carlo method (Table 1) to minimize quality criteria were comparable with the ones obtained by the use of genetic algorithm (Table 2), but the genetic algorithm takes significantly fewer iterations.

Therefore, the genetic algorithm can be successfully applied to the HTS identification process. The genetic algorithm can also be used to optimise the HTS construction, because the construction optimisation uses the same methods of minimization of quality criteria.

Further dissertations for this subject shall study capabilities of more advanced forms of genetic algorithm applications for modelling accuracy improvements.

REFERENCES:

[1] Prokofiev W. N.: Power Engineering and Transport, Moscow: Influence of Interaction of Stream with Limiting Walls for Transient Processes Analysis. 3, 1963, pp. 377-380. (in russian)

[2] Kęsy Z.: Hydrodynamic Torque Converter Control through Properties of Working Fluid, Radom TU Press, Radom, 2003. (in polish)

[3] Kęsy A.: Development of Bladed Wheels with Optimal Parameters for Hydrodynamic Transmission Systems of Transportation Means. DSc Thesis, Radom TU. Radom 1999. (in polish)

[4] Pawelski Z.: Investigation of Hydrodynamic Torque Converter Characteristics for Unsteady State of Load. PhD Thesis. Łódź TU. Łódź 1980. (in polish)

[5] Goldberg D. E.: Genetic Algorithms in Search, Optimization, and Machine Learning. Addison-Wesley Publishing Company, USA 1989.

STUDY OF SAFETY OF HIGH-RISE BUILDINGS USING EVOLUTIONARY SEARCH

S. Khajehpour[1] and D. E. Grierson[2]
[1]*Kinectrics, Generation Plant Technologies, Toronto, Ontario, Canada,*
[2]*Civil Engineering Department, University of Waterloo, Ontario, Canada*

Abstract:　　The paper applies a computer-based method involving evolutionary search and Pareto optimization to investigate the load-path safety of high-rise commercial office buildings against progressive collapse under abnormal loading. The study was motivated by the progressive-collapse failure of the twin towers of the World Trade Center in New York on September 11, 2001. The assessment of load-path safety against progressive collapse is based on the degree of force redundancy that the structural system of a building has. A Pareto-optimal tradeoff surface formed by a population of conceptual designs for a particular office building project is established in the 3D-space of capital cost, operating cost and income revenue. Computer gray-scale filtering of the cost-revenue tradeoff surface is employed to highlight the relative safety of the different building designs. The paper concludes with some general remarks concerning the design of buildings to withstand or delay progressive collapse under abnormal loading.

Keywords:　　high-rise office buildings, load-path safety, progressive collapse, evolutionary search, Pareto optimization

1.　　INTRODUCTION

The tragic failure of the twin towers of the World Trade Center in New York due to terrorist attack on September 11, 2001, will place significant onus on designers of future high-rise marquee buildings to explicitly ensure specified levels of safety against progressive collapse under abnormal loading (impact, blast, fire, etc.). For the particular situation where progressive collapse is triggered by the floor system disengaging from its supports over all or part of the building footprint at a localized story level, as appeared to be the case for the World Trade Center, designers can strive to

T. Burczyński and A. Osyczka (eds),
IUTAM Symposium on Evolutionary Methods in Mechanics, 153–161.
© 2004 Kluwer Academic Publishers.

meet this progressive-collapse safety objective by specifying load-carrying structural systems that have smaller bay areas so as to increase the numbers of girders/columns/shearwalls supporting the floor system. This, and the adoption of floor systems that are well connected to the supporting superstructure, will result in buildings that have high force redundancy and thus enhanced load-path safety against progressive collapse.

This paper investigates load-path safety for an example high-rise office building project. A multi-criteria genetic algorithm is applied to create a number of alternative Pareto-optimal conceptual designs for the building that together form the optimal cost-revenue tradeoff surface in the 3D-space of capital cost, operating cost and income revenue. The calculation of income revenue is based on the premise that larger and more open office space with lots of windows commands a higher annual lease rate. A load-path redundancy function is applied to determine the safety potential of the different building designs. Computer gray-scale filtering of the cost-revenue tradeoff surface is employed to highlight the building concepts having the greatest load-path safety potential against progressive collapse under abnormal loading. Filtering of the tradeoff surface is also used to identify compromise designs having intermediate profit and safety potentials. The work is based upon a computer-based procedure for conceptual design of engineered artifacts developed by the authors [3].

2.   EXAMPLE HIGH-RISING BUILDING

Table 1 lists the parameter values governing an example office building design project [4]. The costs of columns, bracing and shear walls for the building are defined by the unit costs for steel, concrete, reinforcement and forming, while floor and staircase costs are defined by US national averages [5]. The building mechanical and electrical systems costs include those for all-air HVAC systems, electric-traction elevators and fluorescent lighting. The geographical and environmental information apply for a city in North America. The load information is specified by the National Building Code of Canada [6]. Seismic loading is not accounted for. The building architectural systems are specified such that the column lines are regularly spaced in two orthogonal-plan directions. The floor type and depth are taken the same for all stories. The building plan layout, service core area, and floor-to-floor height are specified to be the same for all stories. Table 1 lists the limitations imposed on the building dimensions, as well as on the aspect and slenderness ratios for the building.

Table 1. Governing parameters for office building design

Parameter	Value
Location Information	
Land Unit Cost (US$/m^2)	12000
Range of Annual Lease Rates ($/m^2/yr)	300-540
Maintenance (%capital cost)	2
Taxes (%building value)	5
Mortgage Rate (%)	10
Inflation Rate (%)	3
Unit Costs	
Structural steel ($/ton)	2039
Concrete ($/m^3)	143
Reinforcement ($/ton)	1400
Forming ($/m^2)	45
Roofing ($/m^2)	63
Finishing ($/m^2)	130
Plumbing ($/m^2)	45
HVAC Boiler ($/kW)	225
HVAC Chillers ($/kW)	715
Energy-Electric ($/mWhr)	100
Energy-Gas ($/mWhr)	40
Electrical ($/m^2)	121
Elevators, cladding, windows ($/avgUS$)	1
GEOGRAPHICAL & ORIENTATION INFORMATION	
Latitude (Degree North)	40
Angle of building with East (Degree)	0
ENVIRONMENTAL INFORMATION	
Clear Sky Percentage (%)	75
Hot Day Relative Humidity (%)	80
Cold Day Relative Humidity (%)	50
Inside Temperature (C^0)	22
Average Maximum Outside Temperature (C^0)	31
Average Minimum Outside Temperature (C^0)	□20
Hot Day Temperature Range (C^0)	10
Cold Day Temperature Range (C^0)	10
LOAD INFORMATION	
Applied Dead Load (kN/m^2)	1.45
Gravity Live Load (kN/m^2)	2.80
Wind Load Pressure (kPa)	0.48
Seismic Load	N/A
Building Limitations	
Maximum Footprint Length (m)	70
Maximum Footprint Width (m)	70
Maximum Building Height (m)	300
Minimum Floor/Ceiling Clearance (m)	3
Fixed Core/Footprint Area (%)	20
Minimum Core/Perimeter Distance (m)	7
Minimum Lease Office Space (m^2)	60,000
Maximum Length-to-Width Aspect Ratio	2
Maximum Height-to-Width Slenderness Ratio	9

Table 2 lists the ten choices possible for the load-resisting structural system for the building, along with the different choices possible from among eight floor system types, eight numbers of column bays in either plan direction, sixteen bay widths in either plan direction, four window types, sixteen window ratios, and four exterior cladding types.

Table 2. Primary variables for office building design

Index	Structural System Type	Floor System Type	Bay Number (x , y)	Bay Width (m)	Window Type	Window Ratio (%)	Cladding Type
1	Concrete *m*-frame	Concrete flat plate	3	4.5	Standard	25	PC concrete
2	Concrete *m*-frame & shearwall	Concrete flat slab	4	5.0	Insulated	30	Metal panel
3	Concrete framed tube	Concrete waffle slab	5	5.5	Standard HA	35	Stucco wall
4	Steel *m*-frame	Concrete beam & slab	6	6.0	Insulated HA	40	Glazed panel
5	Steel *g*-frame & bracing	Composite steel beam & concrete slab	7	6.5		45	
6	Steel *m*-frame & bracing	Steel joist & beam & deck & concrete slab	8	7.0		50	
7	Steel *g*-frame & concrete shearwall	Steel beam & composite deck & concrete slab	9	7.5		55	
8	Steel *m*-frame & concrete shearwall	Composite steel beam & deck & concrete slab	10	8.0		60	
9	Steel *g*-frame & bracing & outriggers			8.5		65	
10	Steel framed tube			9.0		70	
11				9.5		75	
12				10.0		80	
13				10.5		85	
14				11.0		90	
15				11.5		95	
16				12.0		100	

g = gravity ; m = moment ; HA = Heat Absorbing ; PC = Pre-Cast

The ranges of available choices for architectural and structural systems listed in Table 2 allow for millions of design concepts for the building, albeit most are infeasible. An evolutionary search technique is employed in the following to identify a subset of feasible designs that are Pareto-optimal in the sense that for each such design there does not exist any other feasible design for the building that simultaneously has smaller capital and operating costs and larger income revenue.

3.   PARETO OPTIMIZATION

A set of optimal feasible design concepts for the building is found by formulating and solving the multi-criteria optimization problem [3]:

Minimize: *{Capital Cost ; Operating Cost ; Income Revenue}*   (1a)

Subject to:*{Dimensional, Availability & Performance Requirements}*   (1b)

In Eq.(1a), the three cost-revenue objective criteria to minimize initial capital cost, annual operating cost and 1/annual income revenue (i.e., maximize annual income revenue) for the office building are functions of the governing parameters and variables for the design listed in Tables 1 and 2. In Eq.(1b), the dimensional restrictions are defined by the building limits listed in Table 1, the availability limitations are defined by the ranges of design variable values listed in Table 2, and the performance requirements ensure that columns, bracing, shear walls and floor systems satisfy design code provisions [1,2] under the action of axial, flexural and shear forces calculated using approximate structural analysis for code-specified combinations of dead, live and wind loading [6].

The Pareto optimization problem posed by Eqs.(1) is solved using a multi-criteria genetic algorithm (MGA), to find Pareto-optimal building designs that are each not dominated for all three cost-revenue objective criteria by any other possible feasible design. The genetic data and operators adopted for the MGA are: population size = 1000 building design concepts (encoded as binary bit-strings); reproduction = weighted roulette-wheel simulation (proportionate fitness selection); crossover = two-point, with 100% probability; and mutation = single-bit, with probability that decreases from 5% to 0% over successive generations of the evolutionary search. The genetic operators are applied generation-after-generation to the population of designs until, guided by cost-revenue fitness evaluations, the Pareto-optimal design set for the building is found to remain the same for a specified number of consecutive generations, at which point the evolutionary search is terminated.

4.   BUILDING DESIGN RESULTS

The MGA evolutionary search finds 815 Pareto designs that together define the optimal cost-revenue tradeoff surface depicted in Figure 1 in the 3D-space of capital cost, operating cost and 1/income revenue for the building [4]; each of the dots corresponds to a different design of the building. Computer gray-scale filtering of the cost-revenue tradeoff surface is carried out in the following to identify zones occupied by building concepts having different load-path safety potential against progressive collapse.

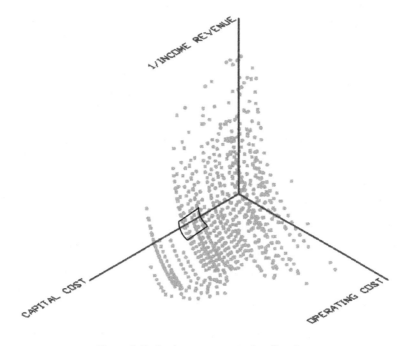

Figure 1. Optimal cost-revenue trade-off surface

Having the numbers of bay areas and columns that define the plan footprint of each Pareto-optimal building design, the corresponding load-path safety potential against progressive collapse triggered by the failure of the entire floor system at a localized story level is assessed by evaluating the force redundancy function

$$R = \{ C * (\text{Number of Bay Areas} + \text{Number of Columns} \ \square 1) \} \qquad (2)$$

where R = the degree of force redundancy (indeterminacy) at any one story level of the building, and C = the degree of force connectivity between the floor system and the girders/columns/shearwalls (e.g., $C = 6$ indicates full bi-axial moment, bi-axial shear, axial and torsional force connectivity of the floor system in all bay areas). The greater the value of R from Eq.(2) the greater is the load-path redundancy of the building and, hence, the greater is its load-path safety potential against progressive collapse under abnormal loading. The relative load-path safety of each building is characterized by a safety index calculated as

$$Safety\ Index = R\ /\ R^{max} \tag{3}$$

where R^{max} is the maximum load-path redundancy from among all buildings in the Pareto-optimal design set. From Eq.(3), the building for which $R = R^{max}$ has the greatest *Safety Index* = 1, while buildings for which $R < R^{max}$ have smaller *Safety Index* < 1.

The gray-scale filtered graphic of the optimal cost-revenue tradeoff surface shown in Figure 2 highlights zones of different load-path safety among the 815 Pareto designs for the building (these results were determined through Eqs. (2) and (3) assuming that the degree of connectivity between the floor system and its supporting girders/columns/shearwalls is the same for all buildings). Figure 2 indicates that but a few buildings have the greatest safety potential (*Zone 4*), the circled one of which is shown in Figure 3. This building has the largest load-path redundancy from among all Pareto designs ($R = R^{max}$) because it has the smallest bay areas (5m x 5m = 25 m^2) and, consequently, proportionally larger numbers of girders and columns available to carry loads.

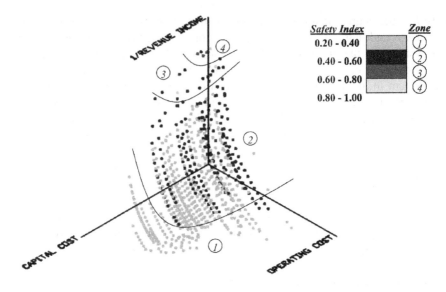

Figure 2. Building load-path safety

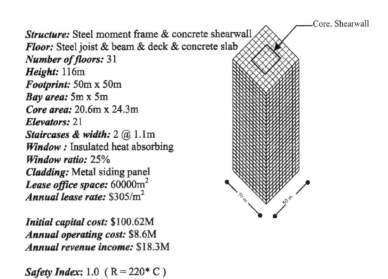

Structure: Steel moment frame & concrete shearwall
Floor: Steel joist & beam & deck & concrete slab
Number of floors: 31
Height: 116m
Footprint: 50m x 50m
Bay area: 5m x 5m
Core area: 20.6m x 24.3m
Elevators: 21
Staircases & width: 2 @ 1.1m
Window : Insulated heat absorbing
Window ratio: 25%
Cladding: Metal siding panel
Lease office space: 60000m²
Annual lease rate: $305/m²

Initial capital cost: $100.62M
Annual operating cost: $8.6M
Annual revenue income: $18.3M

Safety Index: 1.0 (R = 220* C)

Figure 3. Safest building

5. CONCLUDING REMARKS

Examination of the Pareto-optimal design results reveals that the buildings with the greatest safety potential have both lower initial capital cost and lower annual operating cost than buildings with the highest annual income revenue. This is because the floor and façade capital costs and the HVAC operating costs are greater for the latter buildings by virtue of their larger bay areas and window ratios. Moreover, the results also reveal that the load-path redundancy of buildings with higher annual income revenue can be as little as one-third that for the safest buildings; that is, they have somewhat less safety potential against progressive collapse under abnormal loading. Indeed, the safest buildings and those with the greatest annual income revenue represent extremes of the set of designs forming the optimal cost-revenue trade-off surface, in the sense that the highest-revenue buildings have almost the least safety potential against progressive collapse while the safest buildings have almost the least profit potential over time. Perhaps a design that represents a compromise between the two would be a better choice for the building project. In fact, there are multiple alternative Pareto-optimal building designs that have significantly higher safety potential than the highest-revenue buildings and significantly higher profit potential than the safest buildings. To illustrate this, further filtering of the Pareto set identifies the cluster of building designs circumscribed by a box in Figure 1 to have twice the safety and profit potential as the highest-revenue and safest buildings, respectively.

References

[1] CISC 1997: *Handbook of steel construction–seventh edition*. Canadian Institute of Steel Construction, Willowdale, ON, Canada, 1997.

[2] CPCA 1995: *Concrete design handbook – second edition*. Canadian Portland Cement Association, Ottawa, ON, Canada, 1995.

[3] Grierson, D. E.; Khajehpour, S.: Method for conceptual design applied to office buildings. *ASCE J. of Computing in Civil Engineering.* 16, 83-103, 2002.

[4] Khajehpour, S.: *Optimal conceptual design of high-rise office buildings*. PhD Thesis, Civil Engineering, University of Waterloo, ON, Canada, 2001.

[5] Means R.S.: *Assemblies Cost Data / Building Construction Cost Data / Square Foot Costs*, R.S. Means Company, Kingston, MA, USA, 1999.

[6] NRCC 1990: *National Building Code of Canada–NRCC 30619 / Supplement–NRCC 30629*. National Research Council of Canada, Ottawa, ON, Canada, 1990.

STRUCTURAL DESIGN USING GENETIC ALGORITHM

Eisuke Kita
School of Informatics and Sciences
Nagoya University
kita@is.human.nagoya-u.ac.jp

Tatsuhiro Tamaki
Graduate School of Human Informatics
Nagoya University
tamaki@ipl.human.nagoya-u.ac.jp

Hisashi Tanie
Graduate School of Engineering
Nagoya University

Abstract: This paper describes the topology and the shape optimization scheme of continuum structures by using genetic algorithm (GA) and boundary element method (BEM). Structure profile is defined by the help of spline function surfaces. Then, the genetic algorithm is applied for determining the structure profile satisfying the design objectives and the constraint conditions. The present scheme is applied to minimum weight design of two-dimensional elastic problems in order to confirm the validity.

Keywords: topology and shape optimization, genetic algorithm (GA), boundary element method, spline function, two-dimensional elastic problem

1. INTRODUCTION

Many researchers have been studying the application of genetic algorithm to the structural optimization[3, 6]. In the existing studies, structural profiles are represented with cells or function curves. In the cell scheme, the design domain is divided into small square cells. In the function-curve scheme, structural profile is represented with a function

T. Burczyński and A. Osyczka (eds),
IUTAM Symposium on Evolutionary Methods in Mechanics, 163–172.
© 2004 Kluwer Academic Publishers.

curve such as Bezier or spline function. They have some difficulties. In the cell scheme, structures have the zigzag-shaped profiles and therefore, it is necessary to convert the zigzag-shaped profile to smooth one because only the data of smooth profiles are necessary at the design and development processes of the structures. Besides, the number of design variables increases in proportion to the number of cells. When the object is a large-scale structure, the number of design variables becomes numerous. On the other hand, in the function-curve scheme, it is difficult to represent the topology change of the structure.

For overcoming the difficulties, this paper describes the scheme that the boundary profile of a two-dimensional structure is drawn as the profile of the cross-section between the three-dimensional spline surface and a cutting plane which is orthogonal to z-aix. The profiles can be represented by a smaller number of the variables than the cell scheme and, unlike the function-curve scheme, the structural topology can be represented without a special design variable for the topology. Finally, the present scheme is applied to a numerical example.

2.   STRUCTURAL PROFILE REPRESENTATION

The profile representation scheme of the present method is shown in Fig.1.

(a) A system of orthogonal coordinates is taken and lattice points are placed on $x - y$ plane. The coordinates of a lattice point are defined as (X_m, Y_n).

(b) Control points are taken to control the spline surface and the coordinates of a control point are as (X_m, Y_n, Z_{mn}).

(c) A spline surface is formed according to the control points. The coordinates of an arbitrary point on the surface (x, y, z) are given as:

$$\left. \begin{array}{l} x(s,t) = \sum_{m=0}^{M-1}\sum_{n=0}^{N-1} X_m B_{m,k}(s)B_{n,k}(t) \\ y(s,t) = \sum_{m=0}^{M-1}\sum_{n=0}^{N-1} Y_n B_{m,k}(s)B_{n,k}(t) \\ z(s,t) = \sum_{m=0}^{M-1}\sum_{n=0}^{N-1} Z_{mn} B_{m,k}(s)B_{n,k}(t) \end{array} \right\} \quad (1)$$

where s and t denote the variables taken on the x and y-coordinates. The side constraint conditions of s and t are as follows.

$$\left. \begin{array}{l} 0 < s < M-1 \\ 0 < t < N-1 \end{array} \right\} \quad (2)$$

where $B_{\ell,k}$ denotes the B-Spline function. Nodes of the spline function are taken so as to satisfy Scoenberg-Whittny condition.

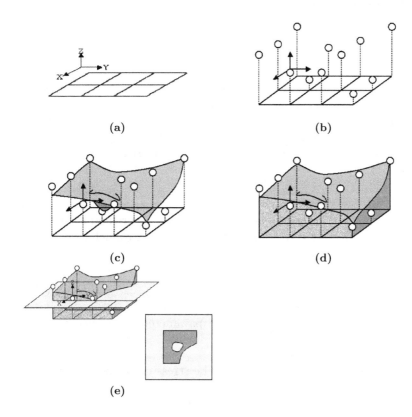

Figure 1. Shape representation scheme

(d) The two-dimensional profile is constructed as the plane between the spline surface and the plane

$$Z = Z_h \tag{3}$$

3. OPTIMIZATION ALGORITHM

3.1 Optimization Problem

The design requirement is to minimize the total weight of the structure on condition that the maximum stress is kept under the permissible stress of the material. The objective function, the constraint condition and the design variables are defined as follows.

3.1.1   Objective Function.   If the material of the structure is kept invariant, the objective function for minimizing the total weight is defined as

$$\frac{A}{A_0} \to \min \qquad (4)$$

where A and A_0 are the area of the structure and the feasible design domain, respectively.

If the objective function is defined from Eq.(4) alone, the final profile may have very strange shape with narrow and long projections. For eliminating such profiles, the following objective function is also considered.

$$\frac{L}{L_0} \to \min \qquad (5)$$

where L and L_0 denote the circumference of the structure and the feasible design domain, respectively. By introducing the parameter C_L, the objective function is defined as follows:

$$f = \frac{A}{A_0} + C_L \frac{L}{L_0} \qquad (6)$$

In the following numerical examples, the parameter C_L is specified to 0.1, which is defined from numerical experiments.

3.1.2   Constraint Condition.   The stress constraint condition is defined as follows:

$$g_0 = \frac{\sigma_{\max}}{\sigma_c} - 1 \geq 0 \qquad (7)$$

where $\sigma_{\max}$ and σ_c denote the maximum value of the von Mises equivalent stress on boundary and the permissible stress of the material, respectively.

Final profile should be constructed with single domain. The related constraint condition is defined as follows:

$$g_1 = N_\Omega - 1 = 0 \qquad (8)$$

where N_Ω means the number of the subdomains. The number of subdomains is counted automatically.

3.1.3   Design Variables.   The z-coordinate of the control point Z_{mn} is selected as design variables. The side constraint for the variable Z_{mn} is defined as

$$Z_0 \leq Z_{mn} \leq Z_1 \qquad (9)$$

where Z_0, and Z_1 denote the minimum and the maximum values.

3.2   Genetic Algorithm[4, 5]

3.2.1   Fitness Function.
The fitness function $fitness$ is defined from the objective function as follows:

$$fitness = 1 - f \tag{10}$$

The constraint conditions are usually added into the fitness function by introducing the penalty parameter. In the present scheme, however, the constraint conditions are not included into the fitness function because the population is organized by the individuals satisfying the constraint conditions alone.

3.2.2   Genetic Coding.
The length of chromosome is equal to the total number of the control points $M \times N$. Each gene of the chromosome is related to the z-coordinate of the control point Z_{mn} as follows:

$$Z_{mn} = Z_0 + (Z_1 - Z_0) \times \frac{G_{m+M \times n}}{G_{\max} - 1} \tag{11}$$

where $G_{\max}$ is the maximum value of the genes.

3.2.3   Genetic Operations.

Selection.   As the selection operator, this paper employs ranking selection, in which the parents are selected from the population according to the ranking of the fitness function of the individuals, instead of the value of the fitness function. Moreover, the elitist scheme is employed so that the best individual at each population survives at the next generation.

Crossover.   One-point crossover is employed in this study.

Mutation.   Mutation operation changes the value of a gene into different value which is selected randomly from arbitrary values.

3.3   Algorithm

Figure 2 shows the algorithm of the present scheme. Firstly, the population is organized by individuals satisfying the constraint conditions alone. The fitness function of each individual is estimated by using boundary element method and then, the genetic operators such as the selection, the crossover and the mutation are applied to the population to create new individuals. New population is organized with the individuals satisfying the constraint conditions. The number of the individuals is kept invariant during the process.

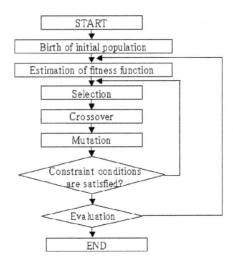

Figure 2. Flowchart of present method

4.   BOUNDARY ELEMENT FORMULATION

We shall explain the BEM analysis in the two-dimensional elastic problem[1, 2].

In two-dimensional elastic body without body forces, the governing equation and the boundary conditions are given as follows:

$$\sigma_{ij,j} = 0 \ \ (\text{in } \Omega) \tag{12}$$

and

$$\left. \begin{array}{l} u_i = \bar{u}_i \ \ (\text{on } \Gamma_u) \\ t_i = \bar{t}_i \ \ (\text{on } \Gamma_t) \end{array} \right\} \tag{13}$$

where u_i, t_i and σ_{ij} denote the displacement, the traction and the stress components in the two-dimensional coordinates, respectively, and $(\ _{,j})$ the derivative in the j direction. Taking the Kelvin solutions as the weight functions, we have the weighted residual equation;

$$\int_\Omega \sigma_{ij,j} u_{ki}^* e_k d\Omega = 0 \tag{14}$$

where e_k denotes the base vectors. Applying the Gauss-Green formula, we have the boundary integral equation;

$$Cu_i = \int_\Gamma (u_{ij}^* t_j - t_{ij}^* u_j) d\Gamma \tag{15}$$

Table 1. Parameters for simulation

Number of control points	$M = N = 8$
Range of Z-coordinates	$Z_0 = 0$, $Z_1 = 1$
Height of a cutting plane	$Z_h = 0.5$
Maximum value of genes	$G_{\max} = 3$
Number of the individuals	100
Crossover rate	0.8
Mutation rate	0.016
Permissible stress	$\sigma_c = 500 kg/mm^2$
Order of Spline function	3

where C is the parameter depending on the placement of the source point; $C = 1$ when the point is inside the domain, $C = 0$ when it is outside the domain and $C = 1/2$ when it is on the smooth boundary. u_{ij}^* and t_{ij}^* denote the fundamental solutions of the displacements and the tractions, respectively. Discretizing Eq.(15) with the boundary elements and collocating the nodes on the elements, we have

$$\boldsymbol{Hu} = \boldsymbol{Gt} \tag{16}$$

where $\boldsymbol{u}$ and $\boldsymbol{t}$ denote the vectors of the nodal displacements and the tractions, respectively. $\boldsymbol{H}$ and $\boldsymbol{G}$ are their coefficient matrices.

Applying the boundary conditions to the above system of equations, we have

$$[\boldsymbol{H}_u \boldsymbol{H}_t] \left\{ \begin{array}{c} \bar{\boldsymbol{u}}_u \\ \boldsymbol{u}_t \end{array} \right\} = [\boldsymbol{G}_u \boldsymbol{G}_t] \left\{ \begin{array}{c} \boldsymbol{t}_u \\ \bar{\boldsymbol{t}}_t \end{array} \right\} \tag{17}$$

where the subscripts u and t denote Γ_u and Γ_t, respectively. Assembling the system of equations, we have

$$[-\boldsymbol{G}_u \boldsymbol{H}_t] \left\{ \begin{array}{c} \boldsymbol{t}_u \\ \boldsymbol{u}_t \end{array} \right\} = [-\boldsymbol{H}_u \boldsymbol{G}_t] \left\{ \begin{array}{c} \bar{\boldsymbol{u}}_u \\ \bar{\boldsymbol{t}}_t \end{array} \right\} \tag{18}$$

and

$$\boldsymbol{Ax} = \boldsymbol{b} \tag{19}$$

The system of equations is solved for $\boldsymbol{x}$ to estimate the unknown nodal values.

5. NUMERICAL EXAMPLE

The object under consideration is shown in Fig.3. The left edge of the object is fixed at the wall and the load of $50 kg/mm^2$ is uniformly given on the upper edge.

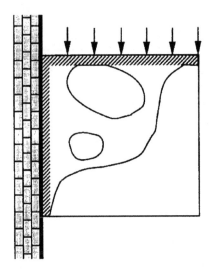

Figure 3.   Object under consideration

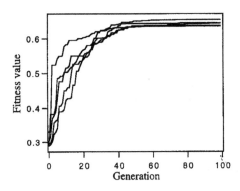

Figure 4.   Fitness value of best individuals

Slash-marked parts of the boundary are fixed during the optimization process because they are the loaded part and the part attached to the wall. The design objective is to minimize the area of the object on the condition that the maximum equivalent stress does not exceed the permissible stress of material. The analysis is carried out according to the parameters shown in Table 1.

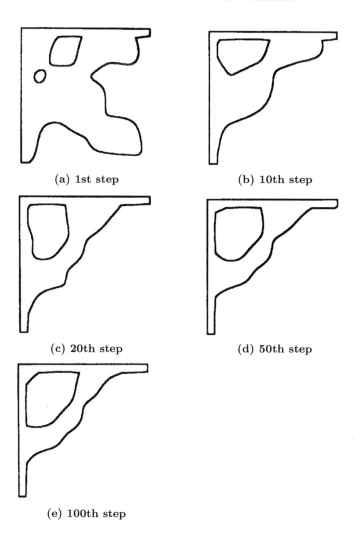

(a) 1st step

(b) 10th step

(c) 20th step

(d) 50th step

(e) 100th step

Figure 5.   Best individuals

Figure 4 shows the convergence properties of the fitness values of best individuals at five runs starting from the different initial population. The abscissa and the ordinate denote the fitness value and the generation, respectively. The fitness values increase as the generation goes and converge to any values at 50th generation. Figure 5 shows the profiles of the best individuals at each generation of a run. The best individual at the initial steps has two holes. Finally, the best individual converges to the profile with one hole as the generation goes.

6.   CONCLUSIONS

This paper presented the topology and shape optimization scheme of the continuum structures by using the genetic algorithm and the boundary element method. The present scheme was applied to a numerical example. The profile of the two-dimensional object is defined as the intersection between the three-dimensional object represented with spline surface and the cutting surface. The minimization of the total weight of the structure and the stress deviation is considered as the objective functions and the coordinates of the controle points of the spline function are taken as the design variables. A final profile satisfying the design objects could be obtained. We may say that validity of basic algorithm of the present method could be confirmed.

References

[1] P. K. Banerjee and R. Butterfield. *Boundary Element Method in Engineering Science*. McGraw-Hill Ltd., 1981.

[2] C. A. Brebbia. *The Boundary Element Method for Engineers*. Pentech Press, 1978.

[3] C. D. Chapman, K. Saitou, and M. J. Jakiela. Genetic algorithms as an approach to configuration and topology design. *J. Mech. Des.*, Vol. 116, pp. 1005–1012, 1994.

[4] L. Davis. *Handbook of Genetic Algorithms*. Van Nostrand Reinhold, 1 edition, 1991.

[5] D. E. Goldberg. *Genetic Algorithms in Search, Optimization and Machine Learning*. Addison Wesley, 1 edition, 1989.

[6] G. Soremekun, Z. Gurdal, R. T. Haftka, and L. T. Watson. Composite laminate design optimization by genetic algorithm with generalized elitist selection. *Computers & Structures*, Vol. 79, pp. 131–143, 2001.

THE TOPOLOGY OPTIMIZATION USING EVOLUTIONARY ALGORITHMS

Grzegorz Kokot, Piotr Orantek
Department for Strength of Materials and Computational Mechanics
Silesian University of Technology, Konarskiego 18a, 44-100 Gliwice, Poland

Abstract: The coupling of modern, alternative optimization methods such as evolutionary algorithms with the effective tool for analysis of mechanical structures - *BEM*, gives a new optimization method, which allows to perform the *generalized shape optimization* (a simultaneous shape and topology optimization) for elastic mechanical structures. This new evolutionary method is free from typical limitations connected with classical optimization methods. In the paper results of researches on the application of evolutionary methods in the domain of mechanics are presented. Numerical examples for some topology optimization problems are presented, too.

Keywords: evolutionary algorithms, genetic algorithms, generalized shape optimization, topology optimization.

1. INTRODUCTION

The application of classical optimization methods is restricted by limitations referring to the continuity of the objective function, the gradient and/or hessian of the objective function and the substantial probability of getting a local optimum. This causes that for some optimization problems the optimal solution is either very difficult or quite impossible to obtain. Therefore new optimization methods, free from limitations mentioned above, have been still looked for. Experiments with mapping natural processes occurring in the nature are promising search directions. Resulting from those experiments various algorithms for searching optimal solutions were created. Those algorithms are known as genetic algorithms,

173

T. Burczyński and A. Osyczka (eds),
IUTAM Symposium on Evolutionary Methods in Mechanics, 173–186.
© *2004 Kluwer Academic Publishers.*

evolutionary programming, evolutionary strategies, neural networks, classifier systems and simulated annealing. Many of them turn out to be alternative methods of optimization for classic methods such as e.g. well known gradient methods. Particularly the genetic algorithms (GAs) are often used in solving optimization problems. They are widely applied in solving the search problems and optimization problems in many disciplines from many years, but their application to solve optimization problems in mechanics has started relatively not long ago.

As the results of the research on application of the genetic algorithms in optimization of mechanical structures, a new approach of the evolutionary optimization has been created. This approach, based on GAs and the boundary element method, is free from limitations connected with classic optimization methods. It allows to solve a large class of optimization problems of mechanical structures concerned with the shape and topology optimization.

2. THE GENERALIZED SHAPE OPTIMIZATION PROBLEM

Consider the following class of optimization problems:

$$\min_{\mathbf{x}} : J_o(\mathbf{x}) \tag{1}$$

with constraints:
$$J_{\square}(\mathbf{x}) = 0, \square = 1,2,\ldots,n, \tag{2}$$

$$J_{\square}(\mathbf{x}) \square 0, \square = 1,2,\ldots,m, \tag{3}$$

$$x_{i_{max}} \square x_i \square x_{i_{min}} \quad i = 1,2,\ldots,k, \tag{4}$$

where $x=(x_i)$ is a vector of design variables.

The functionals $J_o, J_{\square}$ and $J_{\square}$ can have the following forms:

$$J(\mathbf{x}) = \int_{\square} \&(\square,\square,u)d\square + \int_{\square} \%(\mathbf{p},\mathbf{u})d\square \text{ or } J(\mathbf{x}) = \int_{\square} Cd\square \tag{5}$$

where $\&$ is an arbitrary continuous function of stresses $\square$, strains $\square$ and displacements **u** in the domain $\square$ of the structures and $\%$ is an arbitrary function of displacements **u** and tractions **p** on the boundary $\square$.

The functionals (1), (2) and (3) can represent objectives or constraints described by stresses or displacements e.g. in the form of the complementary energy, von Mises stresses, or the cost of the structure. The constraints (4) are simply the geometry constrains.

In the next sections it will be shown how to solve the optimization problem presented above. The proposed method consists of a few steps: geometry modelling and choosing the design variables, applying the numerical method for evaluation of the fitness function, creating the internal voids (if necessary) and applying the evolutionary process. All steps are described below.

3.   GEOMETRY MODELLING

The choice of the geometry modelling method and the design variables has great influence on the final solution of the optimization process. There is a lot of methods for geometry modelling. In the proposed approach NURBS or Bsplines [6] are used to modelling the geometry of the structures. The possibility of the easy control of the shape by only a few control points and local changes of shapes without influence on the rest of the structure are their main advantages (Fig. 1.), which are very helpful in the optimization process.

In the optimization process the co-ordinates of the control points become the design variables, which gives the small number of design variables and the simplicity of data preparation in comparison with other methods (e.g. when the coordinates of boundary nodes (in BEM) or mesh nodes (in FEM) are taken as the design variables).

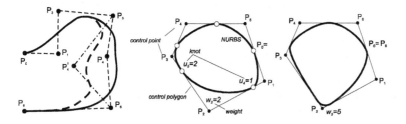

Figure 1. The Bspline and the NURBS

4.   BOUNDARY ELEMENT METHOD

In genetic algorithms the fitness function (the objective function) with constraints plays a role of environment. In the case of optimization problems of mechanical structures the objective function or constraints often depend on stresses, displacements etc. In order to calculate the fitness function values the boundary element method can be used [4].

In the boundary element method the boundary value problem is described by following vector boundary integral equations:

$$c(x)u(x) + \int_\square U^*(x,y)p(y)d\square(y) = \int_\square P^*(x,y)u(y)d\square(y) + \int_\square U^*(x,y)b(y)d\square(y)$$

(6)

where $u(y)$ and $p(y)$ are displacements and tractions on the boundary $\square$, $b(y)$ are the body forces in domain $\square$, $U^*(x,y)$ and $P^*(x,y)$ are fundamental solutions of elastostatics.

Discretizing the boundary $\square$ by means of boundary elements $\square^e$, $e = 1, \dots, E$:

$$x(\ni) = \%_e^w(\ni)(x)_e^w$$

(7)

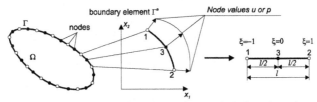

Figure 2. Discretization of the boundary using quadratic boundary elements

Discretization of the boundary using quadratic boundary elements and aproximating the fields of dispacements and tractions on each boundary element $\square^e$, in terms of nodal values and shape functions $\%_e^n$:

$$u(y) = \%_e^n(\ni)u_e^n \qquad p(y) = \%_e^n(\ni)p_e^n$$

(8)

one obtains a discrete form of the boundary integral equation (6):

$$c(x)u(x) = \sum_{e=1}^{E}\sum_{w=1}^{W_e}(u)_e^w \int_{\square^e} P^*[x,y,\ni]\%_e^w(\ni)J(\ni)d\square(\ni) +$$

$$\square \sum_{e=1}^{E}\sum_{w=1}^{W_e}(p)_e^w \int_{\square^e} U^*[x,y,\ni]\%_e^w(\ni)J(\ni)d\square(\ni) + \int_\square U(x,y)b(y)d\square(y)$$

(9)

Finally, equation (9) can be transformed into the system of linear algebraic equations:

$$AX=F$$

(10)

where unknown values of displacements and tractions are described only on the boundary of the structure.

Solving those equations (10) enables obtaining all unknown boundary values of displacements and tractions. Knowing all boundary displacements and tractions on the boundary Γ, the components of the stress tensor $\sigma=(\sigma_{ij})$ can be calculated in selected internal points $x\Omega$ using the following equation:

$$\sigma(\mathbf{x}) = \int_\Gamma D(\mathbf{x},\mathbf{y})\mathbf{p}(\mathbf{y})d\Gamma(\mathbf{y}) - \int_\Gamma S(\mathbf{x},\mathbf{y})\mathbf{u}(\mathbf{y})d\Gamma(\mathbf{y}) + \int_\Omega D(\mathbf{x},\mathbf{y})\mathbf{b}(\mathbf{y})d\Omega(\mathbf{y})$$

(11)

where $S(x,y)$ and $D(x,y)$ are the third-order fundamental solution tensors obtained from suitable differentiation of $U^*(x,y)$ and $P^*(x,y)$ with respect to the source point $\mathbf{x}$ and application of the Hooke's law.

The application of the boundary element method in optimization shows the great advantage over other computational methods for the analysis of mechanical structures, because the discretization is to be done only along the boundary of the structure what reduces the number of data.

5. GENERATING INTERNAL VOIDS, THE BUBBLE METHOD

All methods of solving topology optimization problems are based on the finite element procedures. The boundary element method has found applications to shape optimization problems but till now it has not been used to the topology optimization. In the classical shape optimization one optimizes the shape of the existing boundaries. The boundary element method is the exceptionally natural and convenient numerical technique for such optimization. It appears, however, that it is impossible to insert any changes inside the domain. Because the discretization is made only on the boundary, there is no possibility to insert the void during analysis of the structure.

In order to eliminate this drawback the idea of the coupling of the BEM with the *"bubble method"* [3] has been applied.

It this method it is assumed that inserting an infinitesimally small void into the domain will produce only local stress concentration in the vicinity of the bubble and the global stress field remains unchanged. Looking for the best position of such a bubble, an optimization process is carried out [3], and for some shapes of the bubble (i.e. a circle, a triangle, an ellipse) the so-called *characteristic functions* [3] can be determined:

☐ the circular shaped void

$$H(\sigma_1,\sigma_2) = \frac{1}{2E}\left[(\sigma_1 + \sigma_2)^2 + 2(\sigma_1 + \sigma_2)^2\right] \tag{12}$$

☐ the ellipse-shaped void

$$H = \frac{1}{4E}[5{,}65(\sigma_1 + \sigma_2)^2 + 5{,}52(\sigma_1 - \sigma_2)^2$$
$$- 0{,}22(\sigma_1^2 - \sigma_2^2)\cos 2\sigma + 2{,}34(\sigma_1 - \sigma_2)^2 \cos 2\sigma] \tag{13}$$

☐ the triangle-shape void

$$H = \frac{1}{2E}\left[7{,}0(\sigma_1 + \sigma_2)^2 + 14{,}5(\sigma_1 - \sigma_2)^2\right] \tag{14}$$

The coordinates, where the characteristic function gets the minimum, point out the center of the new generated void. As the characteristic functions depend on stresses it is easy to generate internal voids in the topology optimization process using the boundary element method.

6. EVOLUTIONARY ALGORITHMS

6.1 Genetic algorithms

In general the genetic algorithms (GA) simulate a natural evolutionary process. The genetic algorithms are able to find the optimal solution satisfying the constraints without the calculation of derivatives. Many papers and books contain specimen tests for optimization problems, including ones which are very difficult to solve using classical methods [5].
The genetic algorithms map an evolutionary process of the nature over the span of the age, with the aim to adapt an individual to conditions of life as fit as possible. It is just nothing more than the main goal of optimization. Those algorithms are procedures to search in the feasible space of solutions. They take advantage of mechanisms of natural selection and genetic inheritance, using the neo–Darwinian principle of reproduction and survival of the fittest. At the begining the genetic algorithms work on a population of randomly generated candidates from the feasible solution domain. These candidates, called chromosomes, evolve towards better solutions by applying genetic operators such as selection, mutation, crossover, modeled on the genetic

processes occurring in the nature. After applying genetic operators the new population should have a better fitness. The population undergoes the evolution in a form of the natural selection. An objective function with constraints plays the role of the environment to distinguish between good and bad solutions [4]. In the classical genetic algorithm it is characteristic that the chromosome, representing a possible solution, is described using a binary coding.

6.2 Evolutionary programs

There is a lot of problems for which the GAs can not be applied directly to solve the problem. In such cases there are two possibilities: either the problem is modified in such a way, that the GAs can be applied or GAs is modified in such a way that the problem can be solved.

The first approach allows operating in a domain of the classic genetic algorithm, the second approach introduce in a domain of evolutionary programs. In the presented evolutionary optimization method the modified genetic algorithm, that is evolutionary program, is used.

6.3   Evolutionary optimization based on the bubble method

The flow chart of the proposed approach of the evolutionary optimization is presented in the Fig. 3.

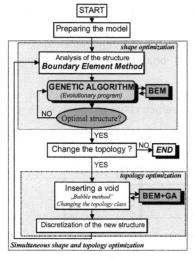

Figure 3. The evolutionary optimization

The first step in optimization process is to find the optimal shape of the external boundary. It means that the typical shape optimization is carried out. As an optimization module the genetic algorithm is applied. In order to obtain the information about the fitness function for each individual in the population, the boundary element method is used. When the main goal of the optimization is to find only the optimal shape without any changes inside the domain, then the optimization process is finished and the obtained solution is the optimal solution.

When it is possible to make any change inside the domain, the second step of the optimization - the topology optimization - is carried on. A new void is inserted into the domain which changes the topology class. In order to generate a new void inside the domain the „bubble method" is used. It secures the optimal position for the inserted void. Coordinates where the characteristic functions get the minimum are the co-ordinates of the center of the new void. The characteristic functions depend on stresses, so minimum values are calculated using the boundary element method and GAs.

After inserting the void, during the optimization process the best place for the inserted void, the optimal shape of the external boundary and the internal boundary are searched. This means that the shape optimization and the topology optimization are carried on simultaneously.

If the "bubble method" is used the first population consists of n the same chromosomes. It makes that the GAs cannot work with a full power in the first stage of evaluation. As the chromosomes in the first population are the same it is not a good "genetic material" for evaluating a best solution.

Just in the next population when the chromosomes are muted and crossed the better "genetic material" for evolution is obtained. But it seems to be a lost time for calculations, because the first population which consists of randomly generated chromosomes from feasible space of solutions looks to be much better for evaluation towards the optimal solution. This new approach is presented in the follows paragraphs.

7.   NUMERICAL EXAMPLES

As the first example a cantilever beam subjected to a point force is considered (Fig. 4.a). The objective function is to minimize the compliance with the constraint condition in the form of the volume constraint ($V_{end} = 0.5V_{start}$). The optimal shape of the external boundary after the shape optimization is presented in Fig. 4.b.

For all presented examples genetic algorithm parameters are as follows:

GA parameters	Value
population size	70
number of generations	300
selective pressure	0.05
probability of nonuniform mutation	2%
probability of crossover	10%

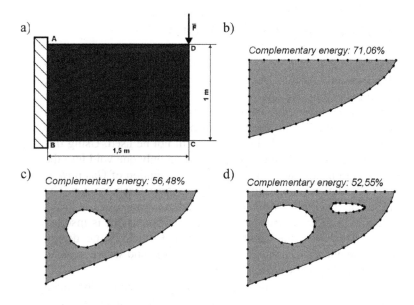

Figure 4. The evolutionary optimization of the cantilever beam

The optimal form of the structure after the simultaneous shape and topology optimization is presented in Fig. 4.c,d.

The second example shows the results of the optimization process of the rectangular plate presented in Fig. 5a. The objective function is to minimize the compliance with the constraint condition in form of the volume constraint (Vend = Vstart). It was assumed that the inserted voids changing the structure topology are circular. The optimal shape of the external boundary is presented in Fig. 5.b. The form of the structure after the simultaneous shape and topology optimization is shown in Fig. 5.c,d.

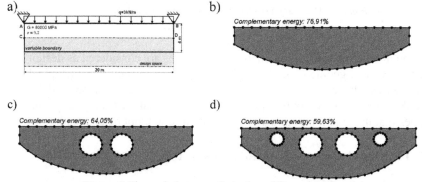

Figure 5. The evolutionary optimization of the rectangular plate

8. ENTIRELY NEW EVOLUTIONARY OPTIMIZATION OF TOPOLOGY

During the numerical tests it was mentioned that there is no necessity to use the „bubble method". As the GAs can start not only from a single point, which is equal to generating the one void, but from population of voids, it is of no importance where the first voids are generated, and it has not an influence on final result. So, in the further work on development of the evolutionary optimization the „bubble method" will not be used. Using this method for generating the new void allowed inserting only one void into the domain in every optimization loop. This led to the limitation of Gas power, because the first populations consisted of nearly the same chromosomes. The new idea is based on generating voids directly by GAs. Then already in the first step a population of different voids (shape and dimension) is inserted into the domain. Moreover, the number of inserted voids is the design variable and GAs decide how many voids should be put and where such an approach was proposed by Burczyński and Orantek [2].

Consider a 2-D elastic structure, which occupies a domain $\square$ bounded by an external boundary $\square$ (Fig. 7). Boundary conditions in the form of a displacement field $u(x)=u_0(x)$, $x\square\square_u$ and a traction field $p(x)=p_0(x)$, $x\square\square_p$, where $\square_u \square \square_p = \square$, are prescribed.

One should find the optimal topology of the structure by introducing a set of voids $\{d_i\}$, $d_i=\{$circle, ellipse, NURBS,...$\}$, i=1,2,...,n_max, into the domain $\square$, to minimize an objective (fitness) function

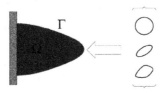

Figure 7. The example of the 2-D elastic structure and their accessible forms of internal voids.

The number of voids, their shape and positions are considered as design variables **X**. The geometrical shape can be imposed in the form of typical plane figures like e.g. circle, ellipse or an arbitrary figure prescribed by NURBS, controlled by control points [6]. For the sake of simplicity one assumes that the external boundary of the structure will not be changed.

8.1 Structures of chromosomes in topology optimization

Three different types of chromosomes are considered:

$$\mathbf{X} = \{X_k\} = < x_1, y_1, \mathbf{r}_1, x_2, y_2, \mathbf{r}_2, ..., x_{n\_max}, x_{n\_max}, \mathbf{r}_{n\_max} > \tag{15}$$

$$\mathbf{X} = \{X_k\} = < n, x_1, y_1, \mathbf{r}_1, x_2, y_2, \mathbf{r}_2, ..., x_{n\_max}, x_{n\_max}, \mathbf{r}_{n\_max} > \tag{16}$$

$$\mathbf{X} = \{X_k\} = < w_1, w_2, ..., w_{n\_max}, x_1, y_1, \mathbf{r}_1, x_2, y_2, \mathbf{r}_2, ..., x_{n\_max}, x_{n\_max}, \mathbf{r}_{n\_max} > \tag{17}$$

where n_max is a maximum number of voids, genes x_i and y_i are coordinates of the center of the *i-th* void and the vector gene, r_i is a vector of shape parameters.

For the circular void the vector gene r_i contains only one element, which is a radius $r_i=\{r_i\}$. For the elliptical void the vector gene r_i contains three elements $r_i=\{r_x, r_y, \square\}$. For more shape complicated voids the vector gene r_i contains a set of radii associated with control points of NURBS $r_i=\{r_1, r_2, ..., r_n\}$.

In the case of the chromosome (15) the number of voids is governed by the condition: if $r_i < r_{min}$ *i-th* void does not exists. It is illustrated for circular voids in Fig. 8.

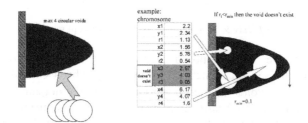

Figure 8. Creating the structural topology on the ground on the chromosome (15).

For the chromosome (16) the number of voids is controlled by the gene n. In the case of the chromosome (17) the number of voids is governed by controlling parameters w_i, which take the value $w_i=0$ if the void does not exist, or $w_i=1$ if the void exists. If $w_i=0$ genes responsible for the shape geometry of i-th void x_i, y_i and r_i are not active.

On each gene, except the controlling genes w_i, there are imposed the following constraints

$$X_k^{min} \square X_k \square X_k^{max} \tag{18}$$

where $X_k^{\min}$ and $X_k^{\max}$ are bounds of variable $X_k^{\max}$.

The flow chart of the evolutionary optimization of the topology is presented in Fig. 8. In order to evaluate the fitness function or constraints one should solve the boundary value problem of theory of elasticity using the finite element method (FEM) or the boundary element method (BEM) [4].

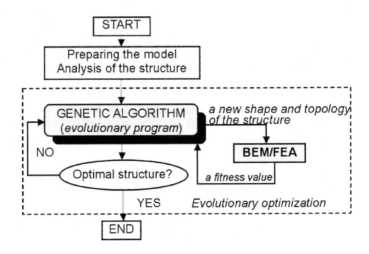

Figure 9. New approach to evolutionary optimization

To examine the proposed evolutionary approach to topology optimization the several numerical examples have been carried out. In all examples circular voids have been introduced into the domain $\square$ to change the topology of the 2-D structures. The maximum number of voids was n_max=4 and the size of the population was 10000. In the evolutionary optimization the same operators of mutations and crossovers were used as in evolutionary identification and the tournament selection was applied.
The chromosomes were constructed according to equation (15).
Three different fitness functions with constraints were applied:

$$f_0 = \int_\square d\square \qquad \text{with constraints:} \quad \square_{eq}\square\square_o \tag{19}$$

$$f_0 = \int_\square \left[\frac{\square_{eq}}{\square_o}\right]^n d\square \qquad \text{with constraints:} \quad |\square\,|\square\square\,_o \tag{20}$$

$$f_0 = \int_{\square} \left[\frac{u}{u_o} \right]^n d\square$$

with constraints: $|\square| |\square\square| o$ (21)

where $\square_o$ - a reference stress, $\square_o$ - an admissible volume of the structure.

Numerical examples of topology optimization are presented in Fig. 10, Fig. 11 and Fig. 12. All received final solutions look like the gruyere cheese structures.

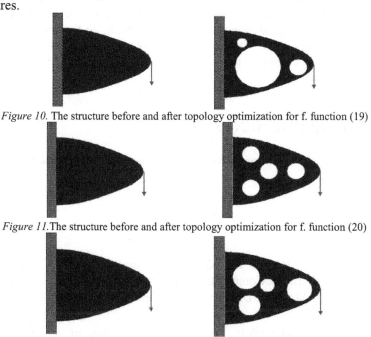

Figure 10. The structure before and after topology optimization for f. function (19)

Figure 11. The structure before and after topology optimization for f. function (20)

Figure 12. The structure before and after topology optimization for f. function (21)

Numerical tests have been also perform for the example presented in the first part of the paper. The results are the same as presented on Fig. 4 and Fig. 5.

9.   CONCLUSIONS

Numerical tests prove that the proposed approach of the evolutionary optimization can be applied to solve a large class of generalized shape optimization problems of mechanical structures (the simultaneous shape and topology optimization). The application of genetic algorithms as the optimization module makes this method free from limitations typical for the classic optimization methods. The coupling of the boundary element method and the genetic algorithms gives an effective and efficient alternative optimization tool.

The application of the boundary element method in optimization shows the great advantage over other computational methods for the analysis of mechanical structures, because the discretization is to be done only along the boundary of the structure. Also the simplicity of data preparation and the little number of data should be taken into consideration.

The proposed approach of the evolutionary optimization enables performing either a complex optimization process in the form of the generalized shape optimization or partial processes in the form of the shape optimization without the change of the topology class or the topology optimization not changing the shape of the external boundary of the structure.

ACKNOWLEDGEMENTS

This research was carried out in the framework of the KBN grant 4T11F00822.

References

[1] Burczyński T., Beluch W., Kokot G., Nowakowski M., Orantek P.: Evolutionary BEM Computation in Optimization and Identification, Proc. IUTAM/IACM/IABEM Symposium, Cracow, 31.5-3.06, 1999, 13-15

[2] Burczyński T., Orantek P., Topology optimization of structures by evolutionary methods. *Proc. Conference on Advanced Materials and Structures*, Warsaw 2000.

[3] Eschenauer H.A., Schumacher A. (1995). Simultaneous shape and topology optimization of structures. WCSMO-1, Proc. on 1st World Congress of Structural and Multidisciplinary Optimization (eds. N. Olhoff, G.I.N. Rozvany), Pergamon, 177-184.

[4] Kleiber M. (Ed.), *Handbook of Computational Solid Mechanics*, Springer, 1998.

[5] Michalewicz Z., *Genetic Algorithms+Data Structures=Evolutionary Programs*, Springer-Verlag, AI Series, New York, (1992).

[6] Piegl L., Tiller W., *The NURBS Book*. Springer Verlag, 1995.

IDENTIFICATION OF THE CMM PARAMETRIC ERRORS BY HIERARCHICAL GENETIC STRATEGY

Joanna Kołodziej

Institute of Mathematics
University of Bielsko-Biała, Bielsko-Biała, Poland
jkolodziej@ath.bielsko.pl

Władysław Jakubiec, Marcin Starczak

Department of Manufacturing Technology and Automation
University of Bielsko-Biała, Bielsko-Biała, Poland
ktmia@ath.bielsko.pl

Robert Schaefer

Institute of Computer Science
Jagiellonian University, Cracow, Poland
schaefer@softlab.ii.uj.edu.pl

Abstract: Hierarchical Genetic Strategy (HGS) is an effective method of global optimization. It performs hierarchical decomposition of admissible domain and utilizes binary strings of different length. We use HGS to solve the problem of identification of the geometrical errors of Coordinate Measuring Machine (CMM). The results of simple numerical experiments will be reported.

Keywords: hierarchical genetic algorithm, global optimization, CMM geometrical errors

1. INTRODUCTION

The following group of problems which take their origin in mechanics may be formulated as the global optimization ones:

- optimal design of structures and mechanical systems (optimal shape design, minimal weight and internal energy of strains, etc.)

T. Burczyński and A. Osyczka (eds),
IUTAM Symposium on Evolutionary Methods in Mechanics, 187–196.
© *2004 Kluwer Academic Publishers.*

- defectoscopy (finding crack and void location, etc.),

- parameter identification in continuous and discrete systems.

Main difficulties in such problems are usually caused by objective non-linearity, large admissible domain, the lack of regularity and global convexity (large plateaus in the objective graph) and existence of many solutions. Such kind of ill posedness exclude the direct application of convex optimization methods and makes the enumerative deterministic or standard stochastic search extremely time consuming. The possible way to pass this obstacles is to apply the refined artificial intelligence methods that can perform robust total search keeping moderate complexity.

Some procedures for evaluation of uncertainty in coordinate measurement require the knowledge of geometrical (translational and rotational) errors of Coordinate Measuring Machine (CMM).The traditional procedures of identification of these errors are time consuming and require the special expansive equipment (see [1,5]). We show how to apply Hierarchical Genetic Strategy (HGS) (see [2,3,4,7]) to deal with these problems.

2.   HIERARCHICAL GENETIC STRATEGY

The main idea of Hierarchical Genetic Strategy (HGS) is running in parallel a set of dependent evolutionary processes. The dependency relation among processes has a tree structure. The processes of low order represent chaotic search with low accuracy. They only detect the promising regions on the optimization landscape in which more accurate processes are activated.

Every single process "builds" a branch of the tree and can be defined as a sequence of evolving populations. The Simple Genetic Algorithm (SGA) is implemented as *the law of evolution* which governs the progression from one generation to the next.

Populations evolving in different branches may consist of individuals with different lengths of genotypes.

We say that the branch has degree $j \in \{1, ..., m\}$ if it is created by populations containing chromosomes of the length $s_j \in \mathbb{N}$.

The unique branch of the lowest degree 1 is called *root* . Populations evolving in this branch contain individuals with genotypes of the shortest length.

We can define a k-periodic metaepoch $M_k, (k \in \mathbb{N})$ as a discrete evolution process which starts from the given population and terminates after at most k generations by selection of the best adapted individual.

Formally, the outcome of the k-periodic metaepoch started from the population $p^{(m)}$ may be denoted by

$$M_k(p^{(m)}) = \left(p^{(m+l)}, \hat{x}, stop\right), l \leq k, \tag{1}$$

where $p^{(m+l)}$ denotes the frequency vector of resulting population, $\hat{x}$ - the best adapted individual in the metaepoch and *stop* - the branch stop criterion flag.

The branch stop criterion detects the lack of progression in the evolution process. It is usually heuristic (e.g. detects the small increment of the average fitness). The branch stop criterion flag is one, if it is satisfied and zero otherwise.

The structure of HGS can be extended by sprouting of the new branches called *children* from the given one by application of the *sprouting operator SO* (see [2,3,4,7]). The branch from which the child is sprouted is called *a parental branch* . We assume, that if the parental branch has degree $j \in \mathbb{N}$, then the degree of its child is $j + 1$.

The sprouting operator can be activated or not, depending on the outcome of prefix comparison operator PC (see [4]).

To start hierarchical algorithm we fix following parameters:

- $1 \leq s_1 < s_2 < s_3 < ... < s_m = s_{max}$ - lengths of binary coded strings ($\Omega_{s_1}, \Omega_{s_2}, ..., \Omega_{s_m}$ are the genetic spaces associated with that strings [2,3],

- $1 \leq n_1 \leq n_2 \leq ... \leq n_m$ - sizes of populations which are the multisubsets of $\Omega_{s_1}, \Omega_{s_2}, ..., \Omega_{s_m}$ respectively,

- $k \in \mathbb{N}$ - a period of metaepoch.

The evolution process starts from the initial population consisting of the shortest binary strings (s_1-length) and creates the root of the structure. The algorithm will evolve creating new branches corresponding to populations containing individuals with different length genotypes.

A process of sprouting of new branch of degree $j + 1$ from the given branch of degree j is called *the branch extending operation* . The detailed definition of this procedure and the whole strategy can be found in [4].

As a *global stop criterion* we can accept a condition when there are no new branches sprouted from the root and the evolution process in every branch is stopped.

3.   MODEL OF GEOMETRICAL ACCURACY OF COORDINATE MEASURING MACHINE (CMM)

The following model for analysis of geometrical accuracy of CMM is used (Fig. 1).

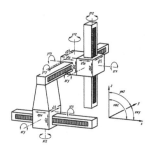

Figure 1.   The model of geometrical errors of coordinate measuring machine.

The designation of geometrical errors of CMM consists of three components: designation of CMM axis concerning the reason of an error, the kind of an error (p - position, t - translation, r- rotation) and the designation of axis on which the error has an influence (refers to the errors of kind t and p) or the designation of rotation axis (refers to the error r).

The components of total error of CMM are the positional parametric errors caused mainly by the errors of scales of each axis: xpx, ypy, zpz. The straightness deviations and the clearance in sliding joints of CMM are the reasons of unintentional linear displacements and rotations. The linear displacements in each axis can be decomposed into two components components in the two planes perpendicular to each other, passing this axis. These errors are so-called translational parametric errors. The x axis has the translational parametric errors xty and xtz, the y axis - ytx and ytz, and the z axis - ztx and zty. The rotations can be expressed as three components for each axis. We call them rotational parametric errors. These components are usually called: roll angle, pitch angle and yaw angle. The x axis has three parametric errors: xrx, xry and xrz, the y axis - yry, yrx and yrz, and finally the z axis - zrz, zrx and zry. The positional, translational and rotational parametric errors (together 18 errors) are the functions of the actual indications x, y and z of linear encoders. In addition there exist three errors of perpendicularity. They are designated as: xwy, xwz and ywz. It means the complete model of CMM parametric errors has 21 components.

The model of the influence of these 21 geometrical parametric errors onto the error of indication for considered CMM can be built. This effect depends on kinematics chain of CMM and the probe configuration. For the CMM being under investigation the error of indication E (in comparison with the location of stylus tip) can be calculated as the product of vector containing the 21 components of geometrical parametric errors and the matrix M which presents the influence (weights) of each of errors onto the x, y and z components of error of indication of CMM:

$$E = KM, \tag{2}$$

where $E = [E_x, E_y, E_z]$ is the vector containing three components of error of indication in specified point in measuring volume and
$K = [ywz, xwz, xwy, ytx, ypy, ytz, yrx, yry, yrz, xpx, xty, xtz,$
$xrx, xry, xrz, ztx, zty, zpz, zrx, zry, zrz]$ is the vector containing 21 components of geometrical parametric errors (18 of them are the functions of indications of linear encoders and remaining 3 are the perpendicularity errors). These parametric errors play the role of the correction coefficients of the machine. The matrix M is given below:

$$
M =
\begin{bmatrix}
0 & -z & 0 \\
-z & 0 & 0 \\
-y & 0 & 0 \\
1 & 0 & 0 \\
0 & 1 & 0 \\
0 & 0 & 1 \\
0 & -z - z_t & y_t \\
z + z_t & 0 & -x - x_t \\
-y_t & x + x_t & 0 \\
1 & 0 & 0 \\
0 & 1 & 0 \\
0 & 0 & 1 \\
0 & -z - z_t & y_t \\
z + z_t & 0 & -x_t \\
-y_t & x_t & 0 \\
1 & 0 & 0 \\
0 & 1 & 0 \\
0 & 0 & 1 \\
0 & -z_t & y_t \\
z_t & 0 & -x_t \\
-y_t & x_t & 0
\end{bmatrix}
$$

where: x, y, z - the indications of CMM linear encoders, x_t, y_t, z_t - the coordinates of center-point of spherical tip of active stylus from from probing system with multiple styli (or the center point of spherical tip for articulated probe in certain location of this probe) in the probe's coordinate system relatively to reference point of probe. Under the assumption that styli or probe location are parallel to CMM global coordinate system, these values are nominally equal to the lengths of styli.

It was assumed that:
- xpx, xty, xtz, xrx, xry, xrz - are the functions of x,
- ypy, ytx, ytz, yry, yrx, yrz - are the functions of y,
- zpz, ztx, zty, zrz, zrx, zry - are the functions of z.

The knowledge of all 21 geometrical parametric errors is necessary to carry-out the mathematical correction of geometrical errors and build the virtual model of CMM.

The measurements of above-mentioned 18 errors are carry-out with the step of digitization about 10-50 mm (depending on the measuring range of CMM and its accuracy). Each measurement is repeated few times for two opposite movements of CMM - the result of measurement is mean value. The measurements of positional, straightness and two rotational errors (pitch and yaw) are carry-out by means of laser interferometer. The measurement of the third rotational error (roll) are carry-out by electronic level (for horizontal axes) or by indirect methods. The number of values describing the CMM accuracy can be calculated using the following formula:

$$\sum_{i=1}^{6} [(L_x/\Delta x_i + 1) + (L_y/\Delta y_i + 1) + (L_z/\Delta z_i + 1)] + 3, \qquad (3)$$

where L_x, L_y, L_z are the measuring ranges of CMM and Δx_i, Δy_i, Δz_i are the digitization steps for evaluation of each of 6 component errors for each axis. For typical CMMs it is about 500-2000 values.

The methodology described above is time-consuming. A lot of research is devoted to finding simpler methods for mathematical correction of CMMs' accuracy. The special attention should be paid to the methods applying the *sphere ball plate* (see [5]).

4.   OPTIMIZATION PROBLEM ORIGINATED BY THE CMM ERRORS IDENTIFICATION

For the sphere ball plate the coordinates of centers of each of 25 balls are known. This coordinates are ordered in Cartesian coordinate system as follows: the center of ball number 1 is the center of the system, the centers of balls 1, 5 and 21 determine the XY plane and the centers of balls number 1 and 5 determine the X axis. The sphere ball plate is placed in certain positions in measuring volume (see [5]).

The coordinates of centers of balls should be transformed in following way: the center of ball number 1 (as in calibration certificate) covers the center of the ball number 1 obtained in measurement results, the center of the ball number 5 (as in calibration certificate) covers with the line defined by the balls number 1 and 5 (as in measurement results) and the center of ball number 21 covers the plane defined by the centers of balls number 1, 5 and 21. Then we should find the values of geometrical errors of CMM (the components of vector $\hat{K}$, which plays the role of optimal correction coefficients that make CMM more accurate), for which the results obtained from virtual model (simulation) differ minimally from the measurement results (in the sense of minimum sum of squares).

We define the objective function F as the superposition of the following operators:

$$F : K \to E \to \left\{ \begin{array}{c} B^i - E = B^i_m \\ i = 1, ..., 25 \end{array} \right\} \to f \qquad (4)$$

where :

- K is the vector containing 21 components of geometrical parametric errors,

- E is the vector containing three components of error of indication in specified point in measuring volume defined by (2),

- $B^i = \begin{bmatrix} x^i \\ y^i \\ z^i \end{bmatrix}$ is the vector of coordinates of the center of ith ball obtained from measurement,

- $B^i_m = \begin{bmatrix} x^i_m \\ y^i_m \\ z^i_m \end{bmatrix}$ is the vector of measured coordinates of the center of ith ball after correction,

-

$$f = \sqrt{\frac{\sum_X \sum_{i=1}^{25} \left[(x^i_m - x^i_t)^2 + (y^i_m - y^i_t)^2 + (z^i_m - z^i_t)^2 \right]}{25(\#X)}} \qquad (5)$$

- (x^i_t, y^i_t, z^i_t) are coordinates of the center of ith ball after transformation,

- X is the discrete set of series of measurements of sphere ball plate in some locations

The value of the objective f has an interpretation of mean error of measurement (in m^{-3}) which remains after corrections of CMM accuracy by using K vector. In other words, we are looking for the vector $\hat{K}$, for which f is minimal.

In order to obtain the maximization problem we put the fitness function:

$$G = \tilde{F} - F, \tag{6}$$

where $\tilde{F}$ is the known upper bound of the possible values of f.

We assume that the coefficients of the positional and translational parametric errors (e.g. xpx, ytz, etc) have the values in range (-0,003; 0,003) and the coefficients of the rotational parametric errors (e.g. yry) have the values in range (-0,000004, 0,000004) and that defines the search space of the problem.

5.   EXPERIMENTAL RESULTS

In our simple experiments we applied classical simple Genetic Algorithm (SGA) and 3-level HGS. The values of parameters for these algorithms are presented in Table 1 below:

Table 1.   Values of parameters for SGA and HGS.

Parameter	SGA	HGS		
		Level 1	Level 2	Level 3
Code length	30	10	20	30
Population size	1000	400	200	50
Mutation rate	0.015	0.03	0.015	0.01
Period of metaepoch	1	10	10	10

The fitness function is conjectured to posses only one global extremum because both algorithms SGA and HGS quickly localize their best individual close one to another: HGS after 10 generation (1 metaepoch) and SGA after about 12 generations. However our main objective was the comparison of both algorithms in the accuracy of the determination of the minimal value of the fitness.

The results of our tests are given in Table 2.

Table 2. The results of experiments for SGA and HGS.

Algorithm	Value of objective	Number of fitness evaluations
SGA	0.0558 after 6000 gen.	240400
HGS	0.0031 after 1700 gen.	148300

Because we know the lower bound of the objective function (it is 0 as usual in case of the identification problem) we can compare both algorithms analyzing the ratio of fitness values corresponding to the best individual. HGS was about 17 times more accurate than SGA. The computational complexity of HGS measured in evaluation of fitness function is also considerably smaller.

The obtained objective value 0,0031 is of the same degree as declared by CMM manufacturer.

6. CONCLUSIONS

- HGS can be a very effective tool in solving ill possed problems of continuous global optimization with multimodal and weakly convex objective functions (see [2,6]). It is also proved for discrete problems with multiple extrema (the inverse kinematic problem in robotics, see [2,3]).

- Ill posedness coming from the lack of objective regularity can be also overcome by HGS. Multilevel HGS can find the accurate solution with moderate complexity without computing objective gradient and Hessian matrix. This is the case of finding very accurate approximation of CMM parametric errors coefficients by 3-level HGS presented in this paper (see Table 2).

- The obtained results show the effectiveness of applied methodology. We hope to obtain better results. In particular, the errors caused by temperature deformations of CMM will be eliminated in our further investigations.

ACKNOWLEDGMENTS

The paper was prepared in the frame of research project no 8 T07D 047 21 supported by KBN (State Committee of Scientific Research)

References

[1] Bosch J.A.(1995): *Coordinate Measuring Machines And Systems,*Marcel Dekker, Inc. New York.

[2] Kołodziej J.(2002): Modeling Hierarchical Genetic Strategy as a Family of Markov Chains, In R. Wyrzykowski, J. Dongarra, M. Paprzycki & J. Waśniewski (eds.),*Parallel Processing and Applied Mathematics,* LNCS, **vol. 2328,** 595-599, Heidelberg : Springer Vlg.

[3] Kołodziej J.(2002):Modeling Hierarchical Genetic Strategy as a Lindenmayer System, *Proc. of International Conference on Parallel Computing in Electrical Engineering PARELEC'2002,* Warsaw, IEEE Computer Society, 409-415, Los Alamitos, CA.

[4] Kołodziej J., Gwizdała R., Wojtusiak J. (2001): Hierarchical Genetic Strategy As A Method Of Improving Search Efficiency. In R. Schaefer and R. Sędziwy (eds.), *Advances in Multi-Agent Systems,* Chapter 9 ,149-161, Cracow : UJ Press.

[5] Kuntzmann H., Trapet E, Waeldele F (1990): A Uniform Concept for Calibration, Acceptance Test and Periodic Inspection of Coordinate Measuring Machines Using Reference Objects. *Annals of the CIRP* , vol. 39/1/1990.

[6] Schaefer R., Kołodziej J. (2002): Genetic search Reinforced by The Population Hierarchy, *Proc. of FOGA VII, A Workshop on Theoretical Aspects of Evolutionary Computation,* Torremolinos, Spain, September 4-6, 2002, Morgan Kaufmann, San Francisco CA.

[7] Schaefer R., Kołodziej J., Gwizdała R., Wojtusiak J. (2000): How Simpletons Can Increase Community Development - an Attempt to Hierarchical Genetic Computation, *Proc. of the 4-th Polish Conference on Evolutionary Algorithms,* Lądek Zdrój, 2000.06.05-08, pp.187-199.

GENETIC ALGORITHM FOR FATIGUE CRACK DETECTION IN TIMOSHENKO BEAM

Marek Krawczuk[1], Magdalena Palacz[2], Wiesław Ostachowicz[3]
[1] Gdańsk University of Technology, Faculty of Electrical and Control Engineering, Narutowicza 11/12, 80-952 Gdańsk, Poland and Institute of Fluid Flow Machinery Polish Academy of Sciences, Fiszera 14, 80-952 Gdańsk, Poland, e-mail:mk@imp.gda.pl,
[2,3] Institute of Fluid Flow Machinery Polish Academy of Sciences, Fiszera 14, 80-952 Gdańsk, Poland, e-mail: [2]mpal@imp.gda.pl, [3]wieslaw@imp.gda.pl

Abstract: Presented paper deals with a method of detecting fatigue cracks in their early state of growth. The method uses optimisation tool consisting of genetic algorithm and gradient method. The applied fitness function is based on the changes in propagating waves.

Keywords: genetic algorithm, crack detection, spectral element method, Timoshenko beam

1. INTRODUCTION

In order to improve the safety, reliability and operational life, it is urgent to monitor the integrity of structural systems. Techniques of non-destructive damage detection in mechanical engineering structures are essential [1], [2], [3]. Previous approaches to non-destructive evaluation of structures to assess their integrity typically involved some form of human interaction. Recent advances in smart materials and structures technology has resulted in a renewed interest in developing advanced self-diagnostic capability for assessing the state of structure without any human interaction. The goal is to reduce human interaction while at the same time monitor the integrity of the structure. With this goal in mind, many researchers have made significant

197

T. Burczyński and A. Osyczka (eds),
IUTAM Symposium on Evolutionary Methods in Mechanics, 197–206.
© 2004 Kluwer Academic Publishers.

strides in developing damage detection methods for structures based on traditional modal analysis techniques. These techniques are often well suited for structures which can be modeled by discrete lumped-parameter elements where the presence of damage leads to some low frequency change in the global behavior of the system [4], [5], [6], [7]. On the other hand small defects such as cracks are obscured by modal approaches since such phenomena are high frequency effects not easily discovered by examining changes in modal mass, stiffness or damping parameters. This is because at high frequency modal structural models are subject to uncertainty. This uncertainty can be reduced by increasing the order of the discrete model, however, this increases the computational effort of modal-based damage detection schemes. There is also a group of methods which utilize thermodynamic damping for assessment of the structural integrity of vibrating structures [8], [9], [10].

Changes in propagating waves are very sensitive to any discontinuities in the structures. Analysis of the process of wave propagation are possible with utilization the spectral element method. This method allows an exact assessment of the inertia of the system [11]. Mathematical site of this method gives only simple set of equations to solve. An important thing is that only one spectral element allows to calculate precisely an infinite number of frequencies and mode shapes of the examined structure. With the spectral element method it is also possible to compare the impact signal and the system response directly in the time domain. Differences in those signals give information about the state of structure.

In the paper a new finite spectral Timoshenko beam element with a transverse open and not propagating crack is introduced. Up till now there are models of a cracked beam in the literature available [11, 12], however there is no spectral model of a cracked Timoshenko one. The crack was modelled with consideration of the influence of the plasticity zone around the crack tip. This approach allows to model the changes of stiffness according to crack appearance in examined structure in more precise way.

For the searching process of parameters of the crack there was used special optimization method, which consisted of simple genetic algorithm and a gradient method. Numerical tests done show that evaluated approach is very sensitive for damage introduction in the structure and the same allows to detect it in very early stage. This fact is very promising for future work in the field of structural health monitoring.

2. MATHEMATICAL MODEL OF THE BEAM

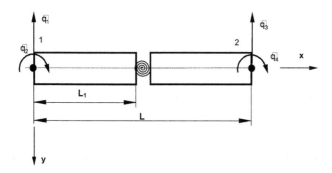

Figure 1. The physical model of the beam.

A spectral Timoshenko beam finite element with a transverse open and non-propagating crack is presented in Fig. 1. The length of the element is L, and its area of cross-section is A. The crack is substituted by a dimensionless and massless spring, whose bending $\square_b$ and shear $\square_s$ flexibilities are calculated using Castigliano's theorem and laws of the fracture mechanics.

Nodal spectral displacements $\hat{w}$ and rotations $\hat{\square}$ are assumed in the following forms, for the left and right part of the Timoshenko beam:

$$\hat{w}_1(x \square (0,L_1)) = R_1 A_1 e^{\square ik_1 x} + R_2 B_1 e^{\square ik_2 x} \square R_1 C_1 e^{\square ik_1 (L_1 \square x)} +$$
$$\square R_2 D_1 e^{\square ik_2 (L_1 \square x)}$$

$$\hat{\square}_1(x \square (0,L_1)) = A_1 e^{\square ik_1 x} + B_1 e^{\square ik_2 x} + C_1 e^{\square ik_1 (L_1 \square x)} + D_1 e^{\square ik_2 (L_1 \square x)}$$

$$\hat{w}_2(x \square (0,L \square L_1)) = R_1 A_2 e^{\square ik_1 (x+L_1)} + R_2 B_2 e^{\square ik_2 (x+L_1)} + \quad (1)$$
$$\square R_1 C_2 e^{\square ik_1 [L \square (L_1 + x)]} \square R_2 D_2 e^{\square ik_1 [L \square (L_1 + x)]}$$

$$\hat{\square}_2(x \square (0,L \square L_1)) = A_2 e^{\square ik_1 (x+L_1)} + B_2 e^{\square ik_2 (x+L_1)} +$$
$$+ C_2 e^{\square ik_1 [L \square (L_1 + x)]} + D_2 e^{\square ik_1 [L \square (L_1 + x)]}$$

where: L_1 denotes the location of the crack, L is the total length of the beam, R_n is the amplitude ratios given by [12]:

$$R_n = \frac{ik_n GAS_1}{GAS_1 k_n^2 \square \square A \square^2} \quad \text{for } (n = 1,2) \quad (2)$$

whereas: $S_1 = \left(\dfrac{0.87 + 1.12\square}{1+\square}\right)^2$ is shear coefficient for displacement [12], $\square$ is Poisson ratio, G is shear modulus, $\square$ denotes density of the material, $\square$ is a frequency and i is imaginary unit given as $i = \sqrt{\square 1}$.

The wave numbers k_1 and k_2 are the roots of the characteristic equation in the general form:

$$(GAS_1 EJ)k^4 \square (GAS_1 \square JK_2 \square^2 + EJ\square A\square^2)k^2 +$$
$$+ (\square JS_2 \square^2 \square GAS_1)\square A\square^2 = 0 \tag{3}$$

where: $S_2 = 12K_1/\ ^2$ is shear coefficient for rotation [12], E denotes Young's modulus and J is second moment of area. The coefficients A_1, B_1, C_1, D_1 A_2, B_2, C_2 and D_2 can be calculated as a function of the nodal spectral displacements using the boundary conditions with additional assumption at the crack place: at the left and right end of the beam displacements and rotations are known, at the crack location transverse displacements, bending moments and shear forces for the left and right part of the beam are the same, whereas the drop in rotations is proportional to the bending moment multiplied by the flexibility of the crack calculated with the fracture mechanics laws [13, 14, 15].

3.　　　FLEXIBILITY AT THE CRACK LOCATION

A coefficients of a beam flexibility matrix at the crack location (in general form) can be calculated using Castigliano theorem [13]:

$$c_{ij} = \frac{\square^2 U}{\square S_i \square S_j} \quad (\text{for } i = 1..6, \ \ j = 1..6) \tag{4}$$

where: U denotes the elastic strain energy of the element caused by the presence of the crack and S are the independent nodal forces acting on the element.

For the analyzed beam, the following relation can express the elastic strain energy due to the crack appearance [14]:

$$U = \frac{1}{E} \int_A (K_I^2 + K_{II}^2) dA, \tag{5}$$

where: A denotes the area of the crack, K_I and K_{II} are a stress intensity factors corresponding to the first and second mode of the crack growth [15].

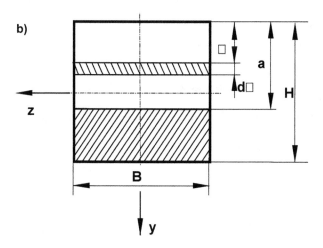

Figure 2. The cross section at the crack location.

The stress intensity factors can be calculated as follows:

$$K_I = \frac{6M}{BH^2}\sqrt{\square}F_I(\frac{\square}{H})$$

$$K_{II} = \frac{\square T}{BH}\sqrt{\square}F_{II}(\frac{\square}{H})$$

(6)

where: M is a bending moment, $\square$ denotes shear factor [16] T is a shear force, B,H $\square$ are dimensions – see Fig.2, F_I and F_{II} are a correction function in the form [15]:

$$F_I\left(\frac{\square}{H}\right) = \sqrt{\frac{\tan(\square/2H)}{\square/2H}}\,\square\frac{0.752 + 2.02(\square/H) + 0.37[1\square\sin(\square/2H)]^3}{\cos(\square/2H)}$$

$$F_{II}\left(\frac{\square}{H}\right) = \frac{1.30\square 0.65(\square/H) + 0.37(\square/H)^2 + 0.28(\square/H)^3}{\sqrt{1\square(\square/H)}}$$

(7)

After simple transformations, the flexibilities of the elastic elements modeling of the cracked cross section of the Timoshenko beam spectral finite element, can be rewritten as:

$$c_b = \frac{72}{BH^2} \int_0^{\bar{a}} \Box F_I^2(\Box)\,d\bar{a}$$

$$c_s = \frac{2\Box}{B} \int_0^{\bar{a}} \Box F_{II}^2(\Box)\,d\bar{a} \tag{8}$$

where: $\bar{a} = \dfrac{a}{H}, \Box = \dfrac{\Box}{H},$ (see Figure.2)

In the non-dimensional form the flexibilities can be expressed as:

$$\Box_b = \frac{EJc_b}{L}$$

$$\Box_s = \frac{GAc_s}{L} \tag{9}$$

4.   OPTIMIZATION METHOD USED

As the genetic algorithm gives only approximate and fully stochastic solution [17] to improve the accuracy of the solution gradient method is implemented into optimization method used. The gradient method utilized was the Newton algorithm for a function with several variables. Gradient method started in two cases: when the mean value of the fitness function reached 95% of the assumed value or when the fitness function value for one "superindividual" reached 97% of the assumed value.

The parameters of the genetic algorithm used were as follows: 30 individuals per population, each individual consisted of 20 bites - 10 bites per variable, fitness function was scaled with linear scaling. Individuals were chosen for recombination with stochastic universal sampling. One-point crossover probability was 90%, mutation probability was 10%.

The fitness function was assumed as:

$$F_f = 1 \Box \left(\frac{|Q_m \Box Q_c|}{|Q_c|} \right)^2 \tag{10}$$

where: Q_m is the "measured" displacement of the beam, Q_c is the displacement calculated for every parameter generated with the genetic algorithm. When the values were the same fitness function was equal to one.

5. EXEMPLARY RESULTS

All numerical calculations were carried out for a steel cantilever beam with geometrical dimensions as follow: length 2 [m], height 0,02 [m], weight 0,02 [m]. The Young's modulus 210 [GPa], mass density 7860 [kg/m^3] and Poisson ratio was 0,3.

Next three figures present results obtained from numerical calculation of the inverse problem. We assumed that we know the size and location of the crack, that means that we calculated the system response for the known values of the crack parameters. Then the optimization method started. To show the advantages of the method there are several results presented.

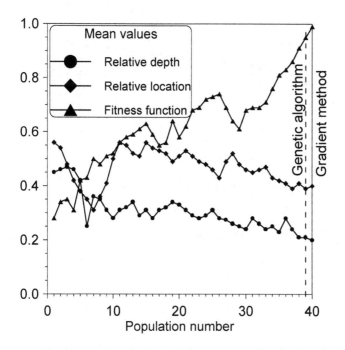

Figure 3. Change of mean values of fitness function, relative crack depth and location in populations.

In the first example (Fig.3) it was assumed that the crack with the depth equal to 20% of the beam height is located in the relative distance from the fixed end equal to 0,4. As the figure shows in the 39th generation there was jump into the gradient method, because the mean value of the fitness function reached 97% of the assumed value. Crack parameters calculated with the genetic algorithm, being the starting point for the gradient method were as follows: the crack depth 0,21 and location 0,39. Values calculated with the gradient method were equal to 0,2 and 0,4, what is the exact value of the searched crack parameters.

In the second numerical example the "searched" relative crack parameters were: the depth 0,15, the location 0,7. According to the figure the gradient method started in the 67th generation, when the mean value of the fitness function reached 97% of the maximum. The gradient method started with values 0,17 and 0,743. The finally calculated values were 0,15 and 0,75. They are exactly the searched crack parameters.

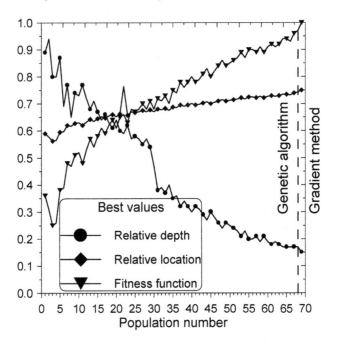

Figure 4. Change of mean values of fitness function, relative crack depth and location in populations.

Last example presented (Fig.5) shows the change of the fitness function and the crack parameters calculated with 53 generations of the genetic algorithm. For this case it was assumed that the crack is located in the relative distance equal to 0,8 and its relative depth is 0,05. The genetic algorithm ended in 53rd generation. Last iteration with values 0,07 and 0,76 was the gradient method. The final calculation values were equal to 0,05 and 0,8, what is the exact match with assumed crack parameters.

As presented above the proposed optimization method allows to find parameters of the crack in a very precise way.

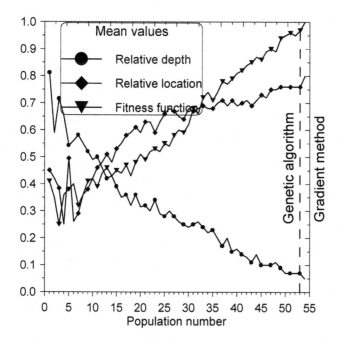

Figure 5. Change of mean values of fitness function, relative crack depth and location in populations.

6. CONCLUSION

The proposed spectral model of a Timoshenko beam with a crack provides analysis of wave propagation in the structure. This fact makes possible utilization of propagating waves for damage detection. Implementing optimization method into the process of searching parameters of the crack saves calculation time and gives precise solution. Proposed method will be developed into elements with more complicated geometry.

References

[1] Structural Health Monitoring: Current Status and Perspectives 1997, edited by Fu-Kuo Chang, Technomic Pub., Basel.
[2] Structural Health Monitoring: Current Status and Perspectives 1999, edited by Fu-Kuo Chang, Technomic Pub., Basel.
[3] Krawczuk M., Ostachowicz W.: 1996, Journal of Theoretical and Applied Mechanics: Damage Indicators for Diagnostic of Fatigue Cracks

in Structures by Vibration Measurements – a Survey. 34(2), pp. 307 – 326.

[4] Adams R.D., Cawley P.: 1979, Journal of Strain Analysis: The Localisation of Defects in Structures from Measurements of Natural Frequencies. 14(2), pp. 49 – 57.

[5] Cawley P., Adams R.D., Pye C.J., Stone B.J.: 1978, Journal of Mechanical Engineering Sciences: A Vibration Technique for Non-Destructively Assessing the Integrity of Structures. 20(2), pp. 93 – 100.

[6] Messina A., Jones I. A., Williams E. J.: 1992, Cambridge, Proceedings of the 1st Conference on Structure Identification: Damage Detection and Localisation Using Natural Frequency Changes, pp. 67 – 76.

[7] Lim T.W., Kashangaki T. A. L.: 1994, AIAA Journal: Structural Damage Detection of Space Truss Structures using Best Achievable Eigenvectors, 30(9), pp. 2310-2317.

[8] Panteliou S. D., Dimarogonas A. D.: 1997, Journal of Sound and Vibration: Thermodynamic Damping in Materials with Ellipsoidal Cavities. 201(5), pp. 555 – 565.

[9] Panteliou S. D., Dimarogonas A. D.: 2000, Theoretical and Applied Fracture Mechanics: The Damping Factor as an Indicator of Crack Severity. 34, pp. 217 – 223.

[10] Panteliou S. D., Chondros T. G., Argyrakis V. C., Dimarogonas A. D.: 2001, Journal of Sound and Vibration: Damping Factor as an Indicator of Crack Severity. 241(2), pp. 235 – 245.

[11] Lakshmanan K. A., Pines D. J.: Modeling Damage in Composite Rotorcraft Flexbeams Using Wave Mechanics, Smart Materials and Structures, 1997, 6, pp. 383 – 392.

[12] Doyle J. F.: Wave Propagation in Structures, Springer – Verlag, New York, 1997.

[13] Przemieniecki J. S.: 1968, New York, McGraw – Hill: Theory of Matrix Structural Analysis.

[14] Papadopoulos C. A., Dimarogonas A. D.: Coupled Longitudinal and Bending Vibrations of a Rotating Shaft with an Open Crack, Journal of Sound and Vibration, 117(1), 1987, pp. 81-93.

[15] Tada H., Paris P. C., Irwin G. R.:, The Stress Analysis of Cracks Handbook, Del Research Corporation, 1973.

[16] Cowper G. R.:, The Shear Coefficient in Timoshenko's Beam Theory, ASME Journal of Applied Mechanics, 1966, pp. 335-340.

[17] Michalewicz Z.: Genetic Algorithms + Data Structures = Evolution Programs, Springer Verlag Berlin, 1996.

MULTICRITERIA DESIGN OPTIMIZATION OF ROBOT GRIPPER MECHANISMS

Stanisław Krenich
Department of Mechanical Engineering. Cracow University of Technology.
31-864 Krakow Al. Jana Pawla II 37, POLAND. Email: krenich@mech.pk.edu.pl

Abstract: In the paper a design optimization problem of robot grippers is formulated. This problem has a multicriteria character in which there are six objective functions and several constraints. To solve the optimization model an evolutionary algorithm based method is proposed. First, the separately attainable minima of the objective functions are found. Next, in order to find the best compromised solution the weighting min–max approach is applied. Finally, the obtained optimal solution is compared with the commercial robot gripper B02 made by Global Modular Gripper (Germany). The comparison of the results shows that using the proposed method we can improve all the criteria.

Keywords: multicriteria design optimization, evolutionary algorithms, robot grippers, robot mechanisms.

1. INTRODUCTION

The problem of optimum design of different mechanisms has quite a long history. Most of these problems are modelled by means of nonlinear programming [1]. In many cases the solution of these models by means of conventional optimization methods might be difficult or even impossible. Thus recently evolutionary algorithms have become an effective tool to solve difficult optimization tasks [4]. The optimization problem of robot grippers has a multicriteria character in which several criteria are to be considered. Thus, the weighting min–max approach is proposed to solve multicriteria design optimization of robot grippers. The optimization problem is to find dimensions of elements of the grippers which satisfy constraints and optimize objective functions.

T. Burczyński and A. Osyczka (eds),
IUTAM Symposium on Evolutionary Methods in Mechanics, 207–218.
© 2004 *Kluwer Academic Publishers.*

2.   PROBLEM FORMULATION

2.1 Optimization Model

Let us consider an example of the mechanism of the commercial robot gripper GMG B04 (Global Modular Gripper-Germany [5]). For this gripper the kinematical scheme is presented in Fig. 1.

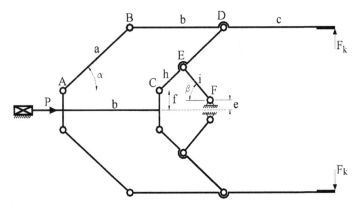

Figure 1. Kinematical scheme of the robot gripper mechanisms.

The simplified optimization model of the robot gripper of this kind was proposed by Krenich & Osyczka in [2]. In this paper additional constraints and objective functions, which include friction forces in the joints, are introduced in the model. These additions make the model closer to the real-life optimization task. The outline of the model can be described as follows:

Vector of decision variables

$\mathbf{x} = [\, a,\, b,\, c,\, e,\, f,\, l,\, h,\, i\,]^{\text{T}}$, where $a,\, b,\, c,\, e,\, f,\, l,\, h,\, i$ are dimensions of the gripper

Objective functions

1. $f_1(\mathbf{x})$ - the difference between maximum and minimum griping forces

$$\text{minimum}\, f_1(\mathbf{x}) = \max_z F_k(\mathbf{x}, z)\,\square\, \min_z F_k(\mathbf{x}, z) \tag{1}$$

where: $\max_z F_k(\mathbf{x}, z),\ \min_z F_k(\mathbf{x}, z)$ are respectively the maximal and minimal gripping force for the assumed range of actuator displacement z, where $Z_{min} < z < Z_{max}$.

2. $f_2(\mathbf{x})$ - the force transmission ratio between the gripper actuator and the gripper ends

$$\text{maximum } f_2(\mathbf{x}) = \frac{\min_{z} F_k(\mathbf{x}, z)}{P}, \tag{2}$$

where: P – the assumed constant actuator force.

3. $f_3(\mathbf{x})$ - the shift transmission ratio between the gripper actuator and the gripper ends

$$\text{maximum } f_3(\mathbf{x}) = \left| \frac{y(\mathbf{x}, Z_{max}) \square y(\mathbf{x}, Z_{min})}{Z_{max} \square Z_{min}} \right| \tag{3}$$

where: $y(\mathbf{x}, Z_{min})$, $y(\mathbf{x}, Z_{max})$– gripper ends displacements at the minimal and maximal position of the actuator end.

4. $f_4(\mathbf{x})$ - the length of all the elements of the gripper

$$\text{minimum } f_4(\mathbf{x}) = \sum_{i=1}^{L} l_i, \tag{4}$$

where: l_i – the length of the i-th element of the gripper mechanism,
L – the number of all the elements of the gripper.

5. $f_5(\mathbf{x})$ - the maximal force in the joints

$$\text{minimum } f_5(\mathbf{x}) = \max_{j} \{R_j\} \text{ for } j=1, 2, ..., LP, \tag{5}$$

where: j – the number of the kinematical joint, LP – the number of all kinematical joints of the gripper mechanism, R_j – the modulus of the force in the j-th kinematical joint.

6. $f_6(\mathbf{x})$ - the mechanical losses in the gripper mechanism

This criterion corresponds with the efficiency of the gripper mechanism.

$$\text{minimum } f_6(\mathbf{x}) = \int_{0}^{Z_{max}} \left(F_k^{BT}(\mathbf{x}, z) \square F_k^{T}(\mathbf{x}, z) \right) dz, \tag{6}$$

where: $F_k^{BT}(\mathbf{x}, z)$– the gripping force for the ideal mechanism without the friction moments in the joints, $F_k^T(\mathbf{x}, z)$ – the gripping force for the real mechanism with the friction in the joints.

Note that the functions $f_1(\mathbf{x})$, $f_4(\mathbf{x})$, $f_5(\mathbf{x})$, $f_6(\mathbf{x})$ are to be minimized, whereas the function $f_2(\mathbf{x})$ and $f_3(\mathbf{x})$ are to be maximized.

Constraints

On the basis of the geometrical dependencies and dependencies between the forces the following geometrical and strength constraints are considered in the model:

Geometrical constraints:

$$g_1(\mathbf{x}) = Y_{min} \square y(\mathbf{x}, Z_{max}) \square 0 \tag{7}$$

$$g_2(\mathbf{x}) = y(\mathbf{x}, z) \square 0 \quad \text{for each position of the actuator} \tag{8}$$

$$g_3(\mathbf{x}) = y(\mathbf{x}, 0) \square Y_{max} \square 0 \tag{9}$$

$$g_4(\mathbf{x}) = Y_G \square y(\mathbf{x}, 0) \square 0 \tag{10}$$

$$g_5(\mathbf{x}) = (h + i)^2 \square (l \square b)^2 \square (f \square e)^2 \square 0 \tag{11}$$

$$g_6(\mathbf{x}) = \frac{1}{2} \square \square > 0, \text{ for each position of the actuator} \tag{12}$$

$$g_7(\mathbf{x}) = \square \square ! \square 0, \text{ for each position of the actuator} \tag{13}$$

$$g_8(\mathbf{x}) = \frac{1}{2} \square \square > 0, \text{ for each position of the actuator} \tag{14}$$

$$g_9(\mathbf{x}) = \square > 0, \text{ for each position of the actuator} \tag{15}$$

$$g_{10}(\mathbf{x}) = c + a \square \cos(\square) \square l + z + b \square Y_{max} \square 0, \text{ for each position of the actuator} \tag{16}$$

Shear stress constraint:

$$g_{11}(\mathbf{x}) = *_{dop} \square \frac{4 \square \max_{j}\{R_j\}}{2 \square \square d^2} \square 0 \tag{17}$$

where $\max_{j}\{R_j\}$ for $j=1, 2, 3, 4, 5, 6$ is the maximal force modulus in the kinematical joints A, B, C, D, E, F.

Minimal gripping force constraint:

$$g_{12}(\mathbf{x}) = F_k(\mathbf{x}, z) \square F_m \square 0 \text{ for each position of the actuator} \tag{18}$$

where:
$y(\mathbf{x}, z)$ – displacement of the gripper ends,
Y_{min} – minimal dimension of the griping object,
Y_{max} – maximal dimension of the griping object,
Y_G – maximal range of the gripper ends displacement,
Z_{max} – maximal displacement of the gripper actuator,
F_m – assumed minimal gripping force,
d – pin diameter in the joints,
$*_{dop}$ - allowable shearing stress,

In addition the upper and lower bounds for the decision variables a, b, c, e, f, h, i, l are used.

2.2 Mathematical dependences in the gripper mechanism

The force dependences:

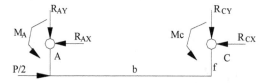

Figure 2. The force distribution in the gripper mechanism – layout 1.

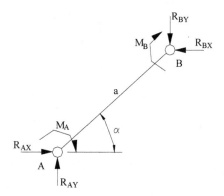

Figure 3. The force distribution in the gripper mechanism – layout 2.

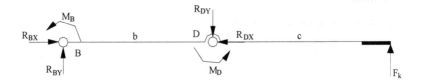

Figure 4. The force distribution in the gripper mechanism – layout 3.

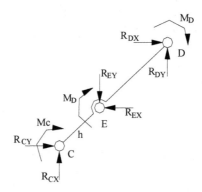

Figure 5. The force distribution in the gripper mechanism – layout 4.

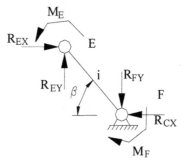

Figure 6. The force distribution in the gripper mechanism – layout 5.

Equations of equilibrium for the layouts presented in figures 2-5 are:

$$\frac{P}{2} \square R_{AX} \square R_{CX} = 0 \tag{19}$$

$$R_{AX} \square R_{BX} = 0 \tag{20}$$

$$R_{AY} - R_{BY} = 0 \tag{21}$$

$$-R_{BX} \cdot a \cdot \sin(\ ) + R_{BY} \cdot a \cdot \cos(\ ) + M_A + M_B = 0 \tag{22}$$

$$R_{BX} - R_{DX} = 0 \tag{23}$$

$$R_{BY} + R_{DY} - F_K = 0 \tag{24}$$

$$-R_{DY} \cdot b + F_K \cdot (b+c) - M_B - M_D = 0 \tag{25}$$

$$R_{CX} + R_{DX} + R_{EX} = 0 \tag{26}$$

$$R_{CY} - R_{DY} + R_{EY} = 0 \tag{27}$$

$$R_{DX} \cdot a \cdot \sin(\ ) + R_{DY} \cdot a \cdot \cos(\ ) + R_{EX} \cdot h \cdot \sin(\ )$$
$$- R_{EY} \cdot h \cdot \cos(\ ) + M_C + M_D + M_E = 0 \tag{28}$$

$$-R_{EX} - R_{FX} = 0 \tag{29}$$

$$-R_{EY} + R_{FY} = 0 \tag{30}$$

$$R_{FX} \cdot i \cdot \sin(\ ) - R_{FY} \cdot i \cdot \cos(\ ) - M_E - M_F = 0 \tag{31}$$

where M_A, M_B, M_C, M_D, M_E, M_F are the friction moments in the kinematical joints A, B, C, D, E, F evaluated from the following formula:

$$M_A = \propto_{cz} \cdot \frac{d}{2} \cdot \sqrt{\left(\frac{P}{2}\right)^2 + R_{AY}^2} \tag{32}$$

$$M_B = \propto_{cz} \cdot \frac{d}{2} \cdot \sqrt{R_{BX}^2 + R_{BY}^2} \tag{33}$$

$$M_C = \propto_{cz} \cdot \frac{d}{2} \cdot \sqrt{R_{CX}^2 + R_{CY}^2} \tag{34}$$

$$M_D = \propto_{cz} \cdot \frac{d}{2} \cdot \sqrt{R_{DX}^2 + R_{DY}^2} \tag{35}$$

$$M_E = \alpha_{cz} \cdot \frac{d}{2} \cdot \sqrt{R_{EX}^2 + R_{EY}^2} \qquad (36)$$

$$M_F = \alpha_{cz} \cdot \frac{d}{2} \cdot \sqrt{R_{FX}^2 + R_{FY}^2} \qquad (37)$$

After transformation the values of the forces in the joints and the gripping force were obtained. The examples of formulas for calculating forces are as follows:

$$R_{AX} = \cdot \frac{P}{2} \cdot \left(\frac{c \cdot h \cdot \sin(\square) \cdot \cos(\square)}{a \cdot b \cdot \sin(\square) \cdot \cos(\square)} + \frac{c \cdot h}{a \cdot b} \right) + M_A \cdot \left(\frac{1}{a \cdot \sin(\square)} + \frac{c}{a \cdot b \cdot \sin(\square)} \right) +$$

$$+ M_B \cdot \left(\frac{1}{a \cdot \sin(\square)} + \frac{c}{a \cdot b \cdot \sin(\square)} + \frac{\cos(\square)}{b \cdot \sin(\square)} \right) + M_C \cdot \left(\frac{c}{a \cdot b \cdot \sin(\square)} \right) +$$

$$M_D \cdot \left(\frac{c}{a \cdot b \cdot \sin(\square)} + \frac{\cos(\square)}{b \cdot \sin(\square)} \right) + M_E \cdot \left(\frac{c \cdot h \cdot \cos(\square)}{a \cdot b \cdot i \cdot \sin(\square) \cdot \cos(\square)} + \frac{c}{a \cdot b \cdot \sin(\square)} \right) +$$

$$+ M_F \cdot \left(\frac{c \cdot h \cdot \cos(\square)}{a \cdot b \cdot i \cdot \sin(\square) \cdot \cos(\square)} \right)$$

$$(38)$$

$$R_{AY} = \cdot \frac{P}{2} \cdot \left(\frac{c \cdot h \cdot \sin(\square)}{a \cdot b \cdot \cos(\square)} + \frac{c \cdot h \cdot \sin(\square)}{a \cdot b \cdot \cos(\square)} \right) + M_A \cdot \left(\frac{c}{a \cdot b \cdot \cos(\square)} \right) +$$

$$+ M_B \cdot \left(\frac{c}{a \cdot b \cdot \cos(\square)} + \frac{1}{b} \right) + M_C \cdot \left(\frac{c}{a \cdot b \cdot \cos(\square)} \right) + M_D \cdot \left(\frac{c}{a \cdot b \cdot \cos(\square)} + \frac{1}{b} \right) +$$

$$+ M_E \cdot \left(\frac{c}{a \cdot b \cdot \cos(\square)} + \frac{c \cdot h}{a \cdot b \cdot i \cdot \cos(\square)} \right) + M_F \cdot \left(\frac{c \cdot h}{a \cdot b \cdot i \cdot \cos(\square)} \right)$$

$$(39)$$

$$R_{BX} = \cdot \frac{P}{2} \cdot \left(\frac{c \cdot h \cdot \sin(\square) \cdot \cos(\square)}{a \cdot b \cdot \sin(\square) \cdot \cos(\square)} + \frac{c \cdot h}{a \cdot b} \right) + M_A \cdot \left(\frac{1}{a \cdot \sin(\square)} + \frac{c}{a \cdot b \cdot \sin(\square)} \right) +$$

$$+ M_B \cdot \left(\frac{1}{a \cdot \sin(\square)} + \frac{c}{a \cdot b \cdot \sin(\square)} + \frac{\cos(\square)}{b \cdot \sin(\square)} \right) + M_C \cdot \left(\frac{c}{a \cdot b \cdot \sin(\square)} \right) +$$

$$+ M_D \cdot \left(\frac{c}{a \cdot b \cdot \sin(\square)} + \frac{\cos(\square)}{b \cdot \sin(\square)} \right) + M_E \cdot \left(\frac{c \cdot h \cdot \cos(\square)}{a \cdot b \cdot i \cdot \sin(\square) \cdot \cos(\square)} + \frac{c}{a \cdot b \cdot \sin(\square)} \right) +$$

$$+ M_F \cdot \left(\frac{c \cdot h \cdot \cos(\square)}{a \cdot b \cdot i \cdot \sin(\square) \cdot \cos(\square)} \right)$$

$$(40)$$

$$R_{BY} = \Box \frac{P}{2} \Box \left(\frac{c \Box h \Box \sin(\Box)}{a \Box b \Box \cos(\Box)} + \frac{c \Box h \Box \sin(\Box)}{a \Box b \Box \cos(\Box)} \right) + M_A \Box \left(\frac{c}{a \Box b \Box \cos(\Box)} \right) +$$

$$+ M_B \Box \left(\frac{c}{a \Box b \Box \cos(\Box)} + \frac{1}{b} \right) + M_C \Box \left(\frac{c}{a \Box b \Box \cos(\Box)} \right) + M_D \Box \left(\frac{c}{a \Box b \Box \cos(\Box)} + \frac{1}{b} \right) +$$

$$+ M_E \Box \left(\frac{c}{a \Box b \Box \cos(\Box)} + \frac{c \Box h}{a \Box b \Box i \Box \cos(\Box)} \right) + M_F \Box \left(\frac{c \Box h}{a \Box b \Box i \Box \cos(\Box)} \right)$$

$$(41)$$

$$F_K = \frac{P}{2} \Box \left(\frac{h \Box \sin(\Box)}{a \Box \cos(\Box)} + \frac{h \Box \sin(\Box)}{a \Box \cos(\Box)} \right) \Box M_A \Box \left(\frac{1}{a \Box \cos(\Box)} \right) \Box M_B \Box \left(\frac{1}{a \Box \cos(\Box)} \right)$$

$$\Box M_C \Box \left(\frac{1}{a \Box \cos(\Box)} \right) + \Box M_D \Box \left(\frac{1}{a \Box \cos(\Box)} \right) \Box M_E \Box \left(\frac{h}{a \Box i \Box \cos(\Box)} + \frac{1}{a \Box \cos(\Box)} \right)$$

$$\Box M_F \Box \left(\frac{h}{a \Box i \Box \cos(\Box)} \right)$$

$$(42)$$

Geometrical dependences in the mechanism:

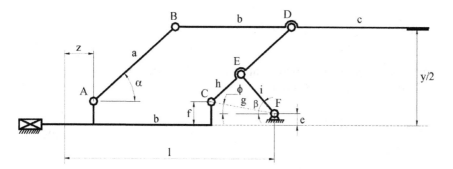

Figure 7. Geometrical dependences in the gripper mechanism.

$$y(\mathbf{x}, z) = 2 \Box \left(f + a \Box \sin(\Box) \right) \qquad (43)$$

$$\Box = \arccos \left(\frac{h^2 + g^2 \Box i^2}{2 \Box h \Box g} \right) \Box ! \qquad (44)$$

$$\Box = \arccos\left(\frac{i^2 + g^2 \Box h^2}{2\Box\Box g}\right) + !$$

(45)

$$! = \arctan\left(\frac{f \Box e}{l \Box z \Box b}\right)$$

(46)

$$g = \sqrt{(f \Box e)^2 + (l \Box z \Box b)^2}$$

(47)

3.   METHODS OF SOLUTION

Traditional optimization methods like gradient based methods, systematic search methods, simplex methods and so on, fail while dealing with the optimization problem of robot grippers. Thus, the constraint tournament selection method is used [3]. This method is used to solve both single and multicriteria optimization problems of robot grippers. Firstly, the separately attainable minima of the objective functions are found. Next, in order to find the best compromised solution the weighting min–max approach is applied. For this approach the preference function is as follows:

$$\mathscr{P}[\mathbf{f}(\mathbf{x})] = \max_{i \Box I}(w_i \Box z_i(\mathbf{x})) \text{ for } i=1, 2,..., 6$$

(48)

where: $z_i(\mathbf{x})$ is the relative increment and w_i is the weighting coefficient for the i-th objective function for which:

$$z_i(\mathbf{x}) = \left|\frac{f_i^0 \Box f_i(\mathbf{x})}{f_i^0}\right|, \quad \sum_{i=1}^{I} w_i = 1$$

(49)

where: f_i^0 is the separately attainable minimum of the i-th objective function.

4.   RESULTS OF OPTIMIZATION PROCESS

Using the constraint tournament method the following separately attainable minima are obtained:

$f_1^0(\mathbf{x}) = 1.185,$ $f_2^0(\mathbf{x}) = 1.399,$ $f_3^0(\mathbf{x}) = 4.167,$

$f_4^0(\mathbf{x}) = 229.58,$ $\qquad f_5^0(\mathbf{x}) = 50.681,$ $\qquad\qquad f_6^0(\mathbf{x}) = 23.16.$

Using the weighting min-max method Pareto optimal solutions for different weighting coefficients are generated. After several tests the following weighting coefficients $w_1=0.07$, $w_2=0.03$, $w_3=0.01$, $w_4=0.87$, $w_5=0.01$, $w_6=0.01$ have given the best compromise solution for which all the objective functions are better than those in the commercial robot gripper. The comparison of the results is presented in Table 1. The force characteristics for the optimal and commercial designs are presented in Fig. 8. For the proposed optimal design this characteristic is made for the integer model.

Table 1. Comparison between obtained optimal gripper mechanisms and the gripperB02-GMG.

		Gripper B02-GMG	Optimal gripper *continuous model*	Optimal gripper *integer model*
Objective functions	$f_1(\mathbf{x})$	917.510	58.372	58.894
	$f_2(\mathbf{x})$	0.312	1.160	1.128
	$f_3(\mathbf{x})$	1.363	2.501	2.487
	$f_4(\mathbf{x})$	380.00	337.995	355.00
	$f_5(\mathbf{x})$	450.415	120.877	121.046
	$f_6(\mathbf{x})$	1370.99	360.55	350.77
Decision variables	a	40.00	33.237	35.0
	b	60.00	35.487	40.0
	c	80.00	120.387	121.0
	e	20.00	10.850	10.0
	f	20.00	15.154	15.0
	h	20.00	13.536	14.0
	i	20.00	18.450	20.0
	l	120.00	90.894	100.0

5. CONCLUSIONS

The results obtained in the paper while optimizing the commercial design of the gripper show the effectiveness of the proposed approach. This means that using the evolutionary based method we can significantly improve the design, i.e., all the assumed criteria can be better than those in the commercial robot gripper. Moreover the evolutionary algorithm, which was used to solve the problem presented above, is a very effective tool to solve highly constrained optimization problems as well as the problems with the computationally expensive objective function. The method reduces the computation time and produce better results.

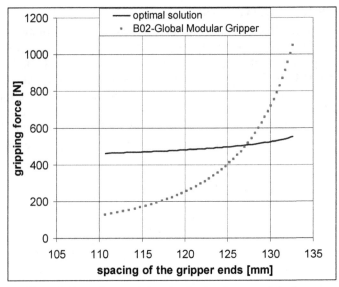

Figure 8. Force characteristics of optimal and commercial grippers

ACKNOWLEDGMENTS

This study was sponsored by Polish Research Committee, under the Grant No. 7 T07A 007 19.

References

[1]   Rao S.S., (1984): Multicriteria Optimization in Mechanisms. Journal of Mechanical Design, Transaction of ASME, 1984, Vol.101, pp. 398-406.
[2]   Krenich S. and Osyczka A. (2000a): Optimization of Robot Grippers Parameters Using Genetic Algorithms. *Proceedings of the 13th CISM-IFToMM Symposium on the Theory and Practice of Robots and Manipulators*, Springer Verlag, Wien, New York, pp.139-146.
[3]   Osyczka A. and Krenich S. (2000b): A New Constraint Tournament Selection Method for Multicriteria Optimization Using Genetic Algorithm. In: *Proc. of the Congress of Evolutionary Computing*, San Diego, USA, pp.501-509.
[4]   Osyczka A. (2002): *Evolutionary Algorithms for Single and Multicriteria Design Optimization*. Springer Physica-Verlag, Heilderberg, Berlin.
[5]   Catalog Global Modular Gripper - GMG, (2000), Gesellschaft fuer Modulare Greifersysteme mbH, Rodinger Weg 8h, D-59494 Soest, htpp://www.gmg-system.com.

OPTIMAL DESIGN OF MULTIPLE CLUTCH BRAKES USING A MULTISTAGE EVOLUTIONARY METHOD

Stanisław Krenich[1] and Andrzej Osyczka[2]
[1]Department of Mechanical Engineering, Cracow University of Technology, 31-864 Krakow Al. Jana Pawla II 37, Poland. email: krenich@mech.pk.edu.pl,
[2]AGH University of Science and Technology, Department of Management, Gramatyka Str. 10, 30-067 Cracow, Poland, email: osyczka@mech.pk.edu.pl

Abstract: In the paper an optimization problem of design of multiple clutch brakes is presented. The problem has a multicriteria character in which there are four objective functions and several constraints. To solve this problem a multistage approach to multicriteria design optimization is proposed. At all stages bicriteria optimization models are solved using evolutionary algorithms. After solving each model the set of Pareto optimal solutions is generated and can be graphically illustrated in the space of objectives. Then this set is presented to the designer who decides on which level one of the objective functions is treated as the constraint. The process is repeated till all the objective functions are considered.

Keywords: multicriteria design optimization, evolutionary algorithms, multiple clutch brakes.

1.   INTRODUCTION

The problem of optimum design of different machine elements and assemblies has quite a long history. Most of these problems are modelled by means of nonlinear programming. In many cases optimization models have a multicriteria character in which using conventional optimization methods to generate a set of Pareto optimal solutions and then to select a compromised solution might be difficult or very limited. Thus recently evolutionary algorithms have become an effective tool for solving multicriteria

T. Burczyński and A. Osyczka (eds),
IUTAM Symposium on Evolutionary Methods in Mechanics, 219–228.
© 2004 Kluwer Academic Publishers.

optimization tasks [1,2,3]. Optimization of multiple clutch brakes is an example of a multicriteria design task in which four criteria are considered. In the paper a multistage evolutionary optimization method is used to solve this task.

2.   OPTIMIZATION MODEL

Let us consider an example of a multiple clutch brake, the configuration of which is shown in Fig.1.

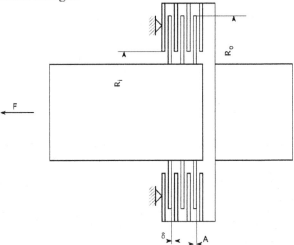

Figure 1. Scheme of the multiple clutch brake

The optimization model of the problem is as follows:
Parameters
The designer must provide the data for the following parameters:

$R_{i\,min}$ = minimum inner radius [mm],

$R_{o\,max}$ = maximum outer radius [mm],

$\Box R$ = minimum difference between radii [mm],

A_{max} = maximum disc thickness [mm],

$\Box$ = distance between discs when unloaded [mm],

L_{max} = maximum length [mm],

Z_{max} = maximum number of discs,

$V_{s\,max}$ = maximum relative speed of the slipstick [m/s],

$\propto$ = coefficient of friction,

$\Box$ = density of material [kg/mm^3],

s = factor of safety,

M_s = static input torque [Nm],

M_f = frictional resistance torque [Nm],

n = input speed [rev/min],

p_{max} = maximum allowable pressure on the disc [Mpa],

J_z = moment of inertia of the working machine [kg mm^2],

t_{max} = maximum stopping time [s],

F_{max} = maximum actuating force [N].

Decision variables

The vector of the decision variables is $\mathbf{x} = [R_i, R_o, A, F, Z]^T$, where:

R_i = inner radius [mm],

R_o = outer radius [mm],

A = thickness of the discs [mm],

F = actuating force [N],

Z = number of friction surfaces.

Objective functions

The vector of the objective functions is:

$$f(\mathbf{x}) = [f_1(\mathbf{x}), f_2(\mathbf{x}), f_3(\mathbf{x}), f_4(\mathbf{x})]^T \tag{1}$$

where:

$f_1(\mathbf{x})$ = mass of the brake [kg],

$f_2(\mathbf{x})$ = stopping time [s],

$f_3(\mathbf{x})$ = outer radius [mm],

$f_4(\mathbf{x})$ = number of friction surfaces [].

All the objective functions are to be minimized and can be evaluated as follows:

$$f_1(\mathbf{x}) = \left(R_o^2 - R_i^2\right) A (Z+1) \tag{2}$$

$$f_2(\mathbf{x}) = t_h = \frac{J_z}{M_h + M_f} \tag{3}$$

$$f_3(\mathbf{x}) = R_o \tag{4}$$

$$f_4(\mathbf{x}) = Z \tag{5}$$

where: braking torque M_h is

$$M_h = \frac{2}{3} \square c \square F \square Z \frac{R_o^3 \square R_i^3}{R_o^2 \square R_i^2} \tag{6}$$

and input angular velocity – is

$$- = \quad \square n / 30 \tag{7}$$

All the objective functions are to be minimized.

Constraints

The inner radius must not be smaller than the specified minimum

$$g_1(\mathbf{x}) \square R_i \square R_{i\min} \square 0 \tag{8}$$

The outer radius must not be greater than the specified maximum

$$g_2(\mathbf{x}) \square R_{o\max} \square R_o \square 0 \tag{9}$$

The distance between the outer and the inner radii must not be smaller than the specified minimum

$$g_3(\mathbf{x}) \square R_o \square R_i \square \square R \square 0 \tag{10}$$

The disc thickness must not be smaller than the specified minimum

$$g_4(\mathbf{x}) \square A \square A_{\max} \square 0 \tag{11}$$

The disc thickness must not be greater than the specified maximum

$$g_5(\mathbf{x}) \square A_{\max} \square A \square 0 \tag{12}$$

The length of the brake must not be greater than the specified maximum

$$g_6(\mathbf{x}) \square L_{\max} \square (Z+1) \square (A+\square) \square 0 \tag{13}$$

The number of friction surfaces must not be greater than the specified maximum

$$g_7(\mathbf{x}) \square Z_{\max} \square (Z+1) \square 0 \tag{14}$$

The number of friction surfaces must not be less than 1

$$g_8(\mathbf{x}) \square Z \square 1 \square 0 \tag{15}$$

The pressure constraint has the form

$$g_9(\mathbf{x}) \equiv p_{max} - p_{rz} \geq 0 \tag{16}$$

where: real pressure p_{rz} is

$$p_{rz} = \frac{F}{S} \tag{17}$$

where: friction surface S is

$$S = \pi \left(R_o^2 - R_i^2 \right) \tag{18}$$

The temperature constraint has the form

$$g_{10}(\mathbf{x}) \equiv p_{max} \, v_{sr\,max} - p_{rz} \, v_{sr} \geq 0 \tag{19}$$

where: the mean relative speed of the slipstick v_{sr} is

$$v_{sr} = \frac{\pi R_{sr} \, n}{30} \tag{20}$$

where: mean friction radius R_{sr} is

$$R_{sr} = \frac{2}{3} \cdot \frac{\left(R_o^3 - R_i^3 \right)}{\left(R_o^2 - R_i^2 \right)} \tag{21}$$

The relative speed of the slipstick constraint has the form

$$g_{11}(\mathbf{x}) \equiv V_{sr\,max} - V_{sr} \geq 0 \tag{22}$$

The stopping time constraint is

$$g_{12}(\mathbf{x}) \equiv t_{max} - t_h \geq 0 \tag{23}$$

The generated torque must be greater than the input torque times the safety factor s

$$g_{13}(\mathbf{x}) \equiv M_h - s \, M_s \, v_{sr} \geq 0 \tag{24}$$

The stopping time must not be less than 0

$$g_{14}(\mathbf{x}) \Box t_h \Box 0 \tag{25}$$

The actuating force must not be less than 0

$$g_{15}(\mathbf{x}) \Box F \Box 0 \tag{26}$$

The actuating force must not be greater than the specified maximum

$$g_{16}(\mathbf{x}) \Box F_{\max} \Box F \Box 0 \tag{27}$$

3.   MULTISTAGE EVOLUTIONARY ALGORITHM BASED METHOD

The decision-making problem is fairly easy when two criteria are considered. This process becomes more difficult when more than two criteria should be considered and when the set of Pareto optimal solutions is large. This occurs while dealing with the optimum design of the multiple clutch brake. For example for the problem presented below 621 Pareto optimal solutions were obtained while using an evolutionary optimization method. Making decision on the bases of this set is a fairly difficult task. Thus in the paper a new multistage optimization process is proposed. The outline of this process is as follows:

Step 1.   Order the objective functions according to their significance for the design process.

Step 2.   Set $n = 1$, where n is the considered stage of the optimization process.

Step 3.   Find the Pareto set for the objective functions $f_n(\mathbf{x})$ and $f_{n+1}(\mathbf{x})$.

Step 4.   Consider the function $f_n(\mathbf{x})$ as the additional constraint of the form $f_n(\mathbf{x}) < F_{mu}$ for minimized functions or $f_n(\mathbf{x}) > F_{nl}$ for maximized functions, where F_{mu} and F_{nl} are suitably the upper and lower restrictions given by the designer.

Step 5.   Set $n = n + 1$, if $n < N-1$, go to Step 2, otherwise go to Step 6.

Step 6.   Check the obtained results and if they are satisfied terminate the calculations, otherwise make a new order of objective functions and repeat the procedure from Step 2.

Verbally this method can be described as follows. At all stages bicriteria optimization models are solved giving in each case Pareto optimal solutions, which can be graphically illustrated in the space of objectives. At each stage from the obtained set of the Pareto optimal solutions, the designer decides how to change one of the two objective functions into a constraint and which new criterion can be considered in the next stage. In particular all decisions of the designer consist in choosing the most preferable ranges of objectives. Note that the results of the optimization process depend on the ordering of the objective functions, i.e., which one is considered as the first objective function, the second and so on. To solve bicriterion optimization models in each step an evolutionary algorithm based method called the constraint tournament selection method [4] is applied.

4.   NUMERICAL EXAMPLE

For the model given above the optimization process was considered as a discrete programming problem. Data for the optimization process were as follows:

Parameters for the brake:

$R_{i\,min}$ = 35 [mm], $R_{o\,max}$ = 110 [mm], $\Box R$ = 20 [mm],

A_{min} = 1.5 [mm], A_{max} = 10.0 [mm], $\Box$ = 0.5 [mm],

L_{max} = 100 [mm], α = 0.5 [ ], $\Box$ = 0.0000078 [kg/mm^3],

p_{dop} = 1 [MPa], Z_{max} = 10, $V_{s\,max}$ = 10 [m/s],

k = 1.8, F_{max} = 1000 [N], t_{max} = 15 [s],

n = 250 [rpm], M_s = 50 [Nm], M_d = 3 [Nm],

I_z = 55 [kg*mm^2].

Data for the evolutionary algorithm:

population size = 200,   number of generations = 400,

crossover rate = 0.6,   mutation rate   = 0.08.

Discrete values of the decision variables:

X_1 = {35, 56, 57, ..., 78, 79, 80}, X_2 = {60, 91, 92,..., 108, 109, 110},

X_3 = {1.5, 2, 2.5, ...,9,5, 10}, X_4 = {600, 610, 620,..., 980, 990, 1000 },

X_5 = {2, 3, 4, 5, 6, 7, 8, 9, 10}.

Assuming that the objective functions are ordered in the same way as in the optimization model, the stages of the optimization process are as follows:

Stage 1

In this stage the following two criteria are considered:

- $f_1(x)$ the function which describes the mass of the brake [kg],
- $f_2(x)$ the function which describes the stopping time [s].

The constraints are given from the basic model by the equations (8-27). After solving the bicriterion model above the set of Pareto optimal solutions generated by the evolutionary algorithm is as presented in Fig.2.

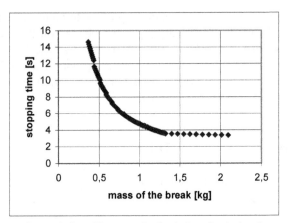

Figure 2. Set of Pareto optimal solutions for the break
design problem from stage 1 (94 solutions).

Stage 2

At this stage the designer decides on which level the first objective
function will be treated as the constraint. Assuming that the mass of the
break should be less than $F_{1u}=1$ [kg], the following constraint is added to the
existing set:

$$g_{17}(\mathbf{x}) = F_{1u} \ \square \ \square\!\left(R_o{}^2 \ \square \ R_i{}^2\right)\square 4 \ \square(Z+1)\square\square\square 0 \tag{28}$$

where: F_{1u} is the assuming upper limit on the first objective function.

The remaining constraints are as considered at Stage 1. A new objective
function is introduced into the optimization model and now the bicriterion
optimization problem is as follows:
- $f_1(\mathbf{x})$ the function which describes the stopping time [s].
- $f_2(\mathbf{x})$ the function which describes the outer radius [mm].

The results of the optimization process for the above model are presented
in Fig.3.

Stage 3

At this stage of the optimization process it is assumed that the stopping time
will be considered as a constraint with the assumed upper bound $F_{2u}=5.5$.
Thus, an additional constraint is added to the model. This constraint has the
form:

$$g_{18}(\mathbf{x}) = F_{2u} \ \square\!\left(\frac{J_z \ \square}{M_h + M_f}\right)\square 0 \tag{29}$$

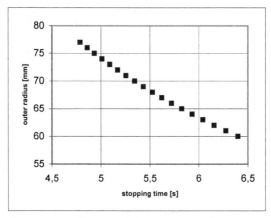

Figure 3. Set of Pareto optimal solutions for the break design
problem from stage 2 (18 solutions).

The remaining constraints are as considered at Stage 2. The objective functions at this stage are:
- $f_1(\mathbf{x})$ the function which describes the outer radius [mm],
- $f_2(\mathbf{x})$ the function which describes the number of friction surfaces.

The results of the optimization process for this stage of calculations are presented in Fig. 4. The solutions from the last set are presented in Table 1. As it is for most multicriteria optimization problems the final decision as to the choice of the solution belongs to the designer. If none of the solutions from the last stage of calculations satisfies the designer, he may repeat calculation from any stage assuming other limit values for the objective function.

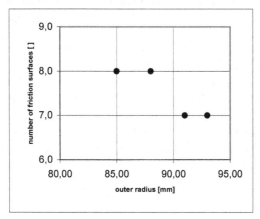

Figure 4. Set of Pareto optimal solutions for the break
design problem from stage 3 (4 solutions).

Table 1. Two solutions from the Pareto set obtained at 3-th stage of the optimization process

	Objective functions				Decision variables				
	$f_1(x)$	$f_2(x)$	$f_3(x)$	$f_4(x)$	R_l	R_o	A	F	Z
1	0.99	4.92	85,0	8,0	65,0	85,0	1,5	980,0	8,0
2	0.97	4.98	93,0	7,0	73,0	93,0	1,5	1000,0	7,0

5.   CONCLUSION

In the paper the multistage process of multicriteria design optimization of multiple clutch brakes is discussed. From the results obtained it is clear that this method provides the designer with a new and very effective tool for solving fairly complicated tasks considering both the complexity of the optimization model and the decision making problem. The method has a universal character and can be applied to solve a wide range of multicriteria design optimization problems.

ACKNOWLEDGMENTS

This study was sponsored by Polish Research Committee (KBN), under the Grant No. 7 T07A 007 19.

References

[1] Coello C.A.C, Veldhuizen D. and Lamont G.: *Evolutionary Algorithms for Solving Multi-Objective Problems*. Kluwer Academic Publisher, New York, 2002.

[2] Deb K.: *Multi-objective Optimization Using evolutionary Algorithms*. Wiley & Sons, Chichester, New York, 2001.

[3] Osyczka A.: *Evolutionary Algorithms for Single and Multicriteria Design Optimization*. Springer Physica-Verlag, Heilderberg, Berlin, 2002.

[4] Osyczka A. and Krenich S.: A New Constraint Tournament Selection Method for Multicriteria Optimization Using Genetic Algorithm. [In:] *Proc. of the Congress of Evolutionary Computing*, San Diego, USA, pp. 501-509, 2000.

DISTRIBUTED EVOLUTIONARY ALGORITHMS IN OPTIMIZATION OF NONLINEAR SOLIDS

W. Kuś[1], T. Burczyński[1,2]

[1] Department for Strength of Materials and Computational Mechanics,
Silesian University of Technology, ul. Konarskiego 18a, 44-100 Gliwice, Poland
[2] Institute of Computer Modelling, Cracow University of Technology.
ul. Warszawska 24, 31-155 Cracow, Poland

Abstract: The paper presents an application of distributed evolutionary algorithms into optimization of nonlinear solids. The distributed evolutionary algorithms work similarly to partially isolated sequential evolutionary algorithms. The advantage of distributed evolutionary algorithms is shorten computational time in comparison with sequential evolutionary algorithms. The fitness function is computed with use of finite (FEM), boundary (BEM) or coupled boundary and finite element methods (BEM-FEM).

Keywords: distributed evolutionary algorithms, optimization, nonlinear, FEM, BEM, coupled BEM-FEM

1.   INTRODUCTION

The shape optimization problem of elasto-plastic structures can be solved using methods based on sensitivity analysis information (e.g. [1][7]) or non gradient methods based on genetic algorithms (e.g. [4][5]). This paper is devoted to the method based on distributed evolutionary algorithms. This work is an extension of previous own papers in which a sequential evolutionary algorithm was used to the shape optimization of elasto-plastic structures [4][5] and present a new method for optimizing structures with geometrical nonlinearities. Applications of evolutionary algorithms in optimization need only information about values of an objective (fitness) function. The fitness function is calculated for each chromosome in each generation by solving boundary - value problem of elasto-plasticity by means of the finite element method (FEM), the boundary element method (BEM) or Coupled Finite and Boundary Element Method (FEM-BEM). This approach does not need information

229

T. Burczyński and A. Osyczka (eds),
IUTAM Symposium on Evolutionary Methods in Mechanics, 229–239.
© *2004 Kluwer Academic Publishers.*

about the gradient of the fitness function and gives the great probability of finding the global optimum. The main drawback of such approach is the long time of calculations. The application of the distributed evolutionary algorithms can shorten the time of calculations but there are additional requirements are need: a multiprocessor computer or a cluster of computers are necessary.

2. FORMULATION OF THE PROBLEM

A body which occupies the domain Ω bounded by the boundary Γ is considered (Fig.2a). The body is made from an elasto-plastic material with hardening. Boundary conditions in the form of displacements and tractions are prescribed and body forces are given. One should find the optimal shape of the body to minimize the areas of plastic zones in the domain Ω. Such an optimization criterion can be achieved by minimizing the fitness function:

$$\min_{\mathbf{x}} F(x) = \int_{\Omega} \left(\frac{\sigma_a}{\sigma_0} \right) d\Omega \qquad (1)$$

where :

$$\sigma_a = \begin{cases} \sigma & \text{when} \quad \sigma \geq \sigma_p \\ 0 & \text{when} \quad \sigma < \sigma_p \end{cases} \qquad (2)$$

σ is a Huber-von Mises yield, σ_0 is a reference stress, σ_p plastic stress and $\mathbf{x}$ is a chromosome vector.

The shape optimization of structures with geometrical nonlinearities could be performed minimizing structure displacements. The fitness function can be formulated in the form:

$$\min_{\mathbf{x}} F(x) = \int_{\Omega} \left(\frac{q}{q_0} \right)^2 d\Omega \qquad (3)$$

where q is the displacement value, q_0 is the reference displacement. Constrains in the form of an admissible volume of the structure and boundary values of design variables are imposed. The shape of the optimized structure is defined using NURBS (Non-Uniform Rational B-Splain)[9]. There is a need of conversing curves into line segments and than the structure is meshed using triangle finite elements (FEM), using boundary elements and cells (BEM), or partially by finite and boundary elements (FEM-BEM). The Triangle [10] code is used for body meshing. The coordinates of NURBS curve control points play the role of genes in the chromosome. Sequentially Genetic Algorithms (SGA) are well known and applied in many areas of optimization problems [8]. The main disadvantage of SGA is the long time needed for computation. The distributed genetic algorithms [11] and distributed evolutionary algorithms (DEA) can make the computation time shorter when more processors are available. The DEA implementation proposed in this paper can be performed on multiprocessor computers or on computers connected by network. DEA work similarly to many sub-evolutionary algorithms logically connected. The global population of chromosomes is divided into sub-populations. Every sub-population evolves separately, and the only connections between sub-populations are done during the migration phase - when some chromosomes migrate between sub-populations. The topology of DEA decides which sub-populations exchange chromosomes between each other. The number of migrating chromosomes and time between migrations are important parameters of DEA (instead of typical SGA parameters). Three types of DEA regarding to a migration model can be considered:

- isolated DEA - there are no migrations,
- synchronous DEA - migrations are produced at the same time,
- asynchronous DEA - migrations are produced at different time, depending on the population state or randomly changed.

In the present research the asynchronous DEA are used, because computers with different speeds were used, and time consumption needed for evaluating the fitness function is different for each chromosome, so synchronizing would slow down the computing. The problem how many migrants should be selected and how frequently a migration should take place are open questions. In the paper the best individual in each sub-population is chosen and migrates into other sub-populations. A new-comer replaces the worst individual in each sub-population. Flowchart of the distributed evolutionary algorithm used in the tests is shown in Fig. 1. An evolutionary algorithm for one sub-population is shown. The evolutionary optimization is performed in a few steps. At first, the starting population is randomly generated. Than a new server thread is created. The server contains chromosomes which emigrate to other sub-populations. The fitness function values for every chromosome are computed using stresses obtained by the direct FEM, BEM or FEM-BEM. The shape of the structure is created by means of genes. Than emigrating chromosomes are selected and copied into servers buffer. If the migration phase occurrs, the client part of application communicates with other sub-evolutionary algorithms and takes chromosomes from theirs buffers. Some chromosomes are put into the population depending on the selection process. The next step is to create the offspring population using evolutionary algorithm operators. The next iteration is performed if the end computing condition is not fulfilled. The end

computing condition can be expressed as the maximum number of iterations. The flowchart of the optimizing algorithm is shown in Fig.1.

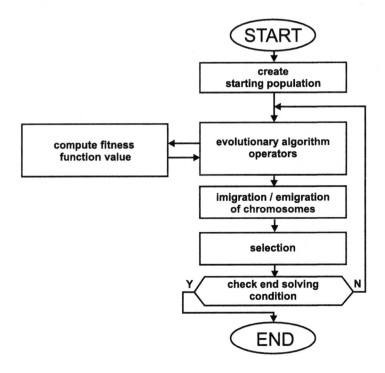

Figure 1.   A distributed evolutionary algorithm (one subpopulation)

The real-coded evolutionary algorithm which operates on chromosomes containing real values was used. Six evolutionary operators were
used: Gauss, uniform and boundary mutation, simple (one point) and
arithmetic crossover and rang selection. Mutation operators introduce
a new genetic material into the population. The uniform and Gauss
mutation changes randomly selected genes in chromosomes adding new
values with uniform or Gauss distribution. Boundary mutation changes
gene value into one of bound values. The crossovers combine the genetic material inside the population. The simple crossover creates a
pair of offspring chromosomes depending on two parent chromosomes.
The offspring contains some genes from one and some from other parent. The arithmetic crossover formats offspring chromosomes containing
genes which values are linear combination of parents. The rank selection
decides which chromosomes will be in the new population. The selection
is done randomly, but the fitter chromosomes have bigger probability to

be in the new population. The probability of being in the new population does not depend on the fitness function value, but on the number of chromosomes ordered according to the fitness function values.

3. COMPUTATION OF THE FITNESS FUNCTION VALUE

3.1 FEM computation

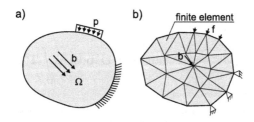

Figure 2. The elasto-plastic body: a) before, b) after FEM discretization

The boundary conditions for an elasto-plastic body (Fig.2a) in the terms of tractions field p and displacement u are prescribed on the boundary Γ. The body forces b are given in the domain Ω. Incremental equilibrium equation can be expressed as follows [6]:

$$\mathbf{K}_T \mathrm{d}\mathbf{q} = \mathrm{d}\mathbf{f} \qquad (4)$$

where $\mathbf{K}_T$ is the tangent stiffness matrix, d$\mathbf{q}$ increment of nodal displacement, d$\mathbf{f}$ increment of nodal forces.

3.2 FEM computation using MSC/Nastran

In some cases (geometrical nonlinearities, shells) MSC/Nastran software is used. Mesh is created on the base of the chromosome genes first. The MSC/Nastran input file is created and MSC/Nastran is invoked to compute a FEM problem. Then output files have to be imported and the fitness function value is computed.

3.3 BEM computation

An elasto-plastic body (Fig.3) is discretized using boundary elements on the boundary and cells inside the body before computing stresses values. The initial strain method is used for solving elasto-plastic structures using the BEM. The boundary integral equation for elasto-plastic

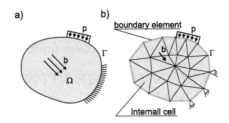

Figure 3.   The elasto-plastic body: a) before, b) after BEM discretization

structures can be expressed as:

$$\mathbf{cu} = \int_\Gamma \mathbf{U}^*\mathbf{p}d\Gamma - \int_\Gamma \mathbf{P}^*\mathbf{u}d\Gamma + \int_\Omega \mathbf{U}^*\mathbf{b}d\Omega + \int_\Omega \mathbf{T}^*\varepsilon^p d\Omega \qquad (5)$$

where $\mathbf{U}^*$, $\mathbf{P}^*$, $\mathbf{T}^*$ are fundamental solutions, $\mathbf{u}$ is an displacement vector, $\mathbf{p}$ is the tractions vector, b is body loads vector and ε_p are plastic strains. The stresses inside the body can be achieved using the equation:

$$\sigma(\mathbf{x}) = \int_\Gamma \mathbf{D}^*\mathbf{p}d\Gamma - \int_\Gamma \mathbf{S}^*\mathbf{u}d\Gamma + \int_\Omega \mathbf{D}^*\mathbf{b}d\Omega + \int_\Omega \mathbf{T}^*\varepsilon^p d\Omega + \mathbf{I}^\sigma \varepsilon^p(x) \quad (6)$$

where $\mathbf{D}^*$, $\mathbf{S}^*$ are fundamental solutions and $\mathbf{I}^\sigma$ is a coefficients matrix.

3.4   Coupled FEM-BEM computation

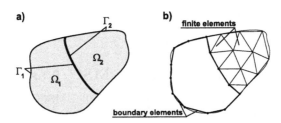

Figure 4.   The elasto-plastic body: a) before, b) after FEM-BEM discretization

The domain of a body Ω is divided into the BEM Ω_1 and the FEM Ω_2 regions. The region Ω_1 has the boundary Γ_1 and the region Ω_2 - the boundary Γ_2, as shown in Fig. 4. Nodes (but not elements) on the intersected part of the boundaries are common for the BEM and the FEM. The FEM part of the body has to be discretized using finite elements and the BEM part using boundary elements (one assumes that body forces in BEM region do not exist). For the FEM part of the body

one can write:
$$\mathbf{K}^1\mathbf{u} = \mathbf{f}^1 \qquad (7)$$

One can write equation for the BEM region without plastic strains as:
$$\mathbf{Hu} = \mathbf{Gp} \qquad (8)$$

where $\mathbf{H}$ and $\mathbf{G}$ are coefficient matrices, $\mathbf{u}$ is the displacements vector, $\mathbf{p}$ is the tractions vector. After rearrangement one achieves:
$$\mathbf{MG}^{-1}\mathbf{Hu} = \mathbf{M}p \qquad (9)$$

where $\mathbf{M}$ is a shape function matrix. Substituting $\mathbf{K}^2 = \mathbf{MG}^{-1}\mathbf{H}$ and $\mathbf{f}^2 = \mathbf{M}p$ one obtains:
$$\mathbf{K}^2\mathbf{u} = \mathbf{f}^2 \qquad (10)$$

which is similar to the FEM system of equations. Matrices $\mathbf{K}^1$, $\mathbf{K}^2$ and vectors $\mathbf{f}^1$, $\mathbf{f}^2$ can be agregated. The final equation is computed iteratively due to nonlinearities in the FEM region. The matrix $\mathbf{K}^1$ is substituted by the tangent stiffness matrix $\mathbf{K}^1_T$.

4. NUMERICAL EXAMPLES

In each test problem a material with the characteristic presented in Fig. 5 is used. E_1 and E_2 are Young's modules, ε_p is the yield strain

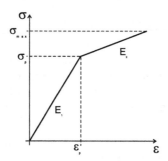

Figure 5. Uniaxial stress-strain curve for material used in tests

and σ_p is the yield stress.

Numerical example 1.

A 2-D structural element is considered (Fig.6a). The external boundary and the hole boundary undergo shape optimization. The external boundary was modelled using the NURBS curve with 3 control points (one of them can be moved - 2 design variables) and the internal hole was modelled using 4 control points NURBS curve (each can be moved - 8 design variables). The fitness function was computed using the FEM. The material data are: $E_1 = 2 \cdot 10^5 MPa$, $E_2 = 0.5 \cdot 10^5 MPa$, $\sigma_p = 250MPa$, $\mu = 0.3$, plate thickness $5mm$. The parameters of the distributed evolutionary algorithm are: the number of subpopulations - 4, the number of chromosomes in subpopulations - 125, migration occurs in every generation, the number of emigrating chromosomes - 1. The shape of the boundary after the first and the 196th generation is shown in Fig. 6. The plastic areas are marked using gray color.

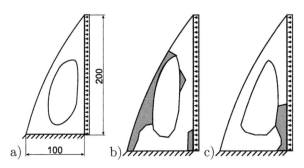

Figure 6. The optimized plate: a) geometry, b) best after 1st generation, c) best after 196th generation

Numerical example 2.

The problem of the shape optimization of a half K-structure is considered (Fig.7). The material data and parameters of the DEA are given in Table 3. The traction-free boundary is modelled by 2 NURBS curves with 3 control points each. The fitness function was evaluated by the BEM. The material data are: $E_1 = 2 \cdot 20^5 MPa$, $E_2 = 0.5 \cdot 10^5 MPa$, $\sigma_p = 150MPa$, $\mu = 0.3$, plane strain. The parameters of the distributed evolutionary algorithm are: the number of subpopulations - 4, the number of chromosomes in subpopulations - 100, a migration occurs in every generation, the number of emigrating chromosomes - 1. The shape of the structure after first and the 476th generation is shown in Fig.7. Gray color was used to mark the plastic areas.

Numerical example 3.

A 3-D shell is considered (Fig.8a). The shell has 10 holes with the constant radius. The holes can be moved. The optimization criterion is to minimize integral over shells displacements. The shell was computed considering large displacements. The fitness function was evaluated using MSC/Nastran. The material data are: $E_1 = 2 \cdot 10^5 MPa$, $E_2 = 0.5 \cdot 10^5 MPa$, $\sigma_p = 300MPa$, $\mu = 0.3$, plate thickness $10mm$. The parameters of the distributed evolutionary algorithm are: the number of subpopulations - 4, the number of chromosomes in subpopulations - 250, migration occurs in every generation, the number of emigrating

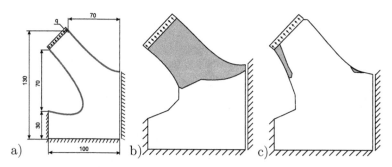

Figure 7. A half of K-structure: a) geometry, b) best after 1st generation, c) best after 476th generation

chromosomes - 1. The shape of the shell after the first and the 500th generation is shown in Fig.8.

Figure 8. Shell: a) geometry, b) best after 1st generation, c) best after 500th generation

Numerical example 4.

Infinite domain with 3 holes is considered (Fig.9a). The holes have constant radii. There are 6 design variables, each pair of variables defines the center position of the hole. The holes could be placed inside the region marked with doted line (Fig.9a). The region near the holes is discretized using finite elements and the rest of the region is discretized using boundary elements. The optimization goal is to minimize the areas of plastic strains, changing the positions of the holes. Constrains on the variability of hole center positions are imposed. The material data are: $E_1 = 2 \cdot 10^5 MPa$, $E_2 = 0.5 \cdot 10^5 MPa$, $\sigma_p = 300 MPa$, $\mu = 0.3$, plate thickness $1mm$. The parameters of the distributed evolutionary algorithm are: the number of subpopulations - $1-4$, the number of chromosomes in subpopulations - 30, a migration occurs in every generation, the number of emigrating chromosomes - 10. The optimal distribution of holes for best chromosome in first and the last generation are shown in Fig. 9.

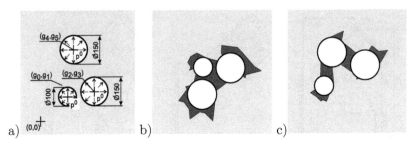

Figure 9. Holes in infinite medium: a) geometry, b) best after 1st generation, c) best after optimization

The speedup achieved using the distributed evolutionary algorithm with different numbers of processors were measured The problem was solved several times for different number of processors. In the tests two, two-processors 1.4GHz Pentium 4 computers were used. The tests were interrupted after achieving the assumed fitness function value. The speedup of the optimization is shown in Fig. 10. The speedup is defined as a time needed to perform the optimization on single processor divided by the time needed on n-processors. The linear speed up which is the maximal speed up for the parallel evolutionary algorithm is also marked in Fig. 10.

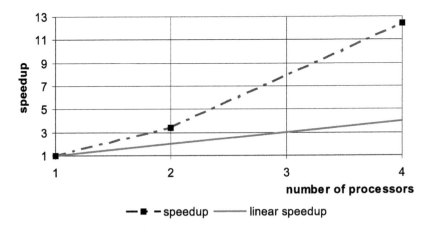

Figure 10. Speedup of computations

5. CONCLUDING REMARKS

Evolutionary shape optimization of nonlinear structures is as a rule the time-consuming because the nonlinear boundary-value problem should be solved for each chromosome. Applications of the DEA considerably speed up the evolutionary process of optimization.

ACKNOWLEDGMENTS

This research was carried out in the framework of the KBN grant 4T11F00822.

References

[1] Burczyński T., Applications of BEM in sensitivity analysis and optimization *Computational Mechanics*, **13**, No. 1/2, 1993.

[2] Burczyński T., Recent Advances in Boundary Element Approach to Design Sensitivity Analysis - Survey, Chapter 1. *Design Sensitivity Analysis* (Eds. M.Kleiber and T.Hisada), Atlanta Technology Publications, Atlanta, 1993.

[3] T. Burczyński, W.Beluch, A.Dugosz, P.Orantek and M.Nowakowski, Evolutionary methods in inverse problems of engineering mechanics, Proc. ISIP 2000, Nagano, 2000.

[4] T. Burczyński, W. Beluch, G. Kokot, Optimization of cracked structures using boundary elements and evolutionary computation, In: *Boundary Element Techniques* (ed. M.H.Aliabadi), London, 1999.

[5] T. Burczyński, A. Długosz, Evolutionary Optimization in Thermoelastic Problems using the Boundary Element Method. Proc. Symposium of the International Association for Boundary Element Methods, Brescia Italy, 2000.

[6] T. Burczyński, J.H. Kane, and C. Balakrishna, Comparison of shape design sensitivity analysis via material derivative-adjoint variable and implicit differentiation techniques for 3-D and 2-D curved boundary elements, *Computer Methods in Applied Mechanics and Engineering*, **142**, 1997, pp.89-109, 1997.

[7] T. Burczyński and W. Kuś, Evolutionary methods in shape optimization of elastoplastic structures 33rd Solid Mechanics, Zakopane, 2000.

[8] T. Burczyśki, G. Kokot, The evolutionary optimization using genetic algorithms and boundary elements, Proc. 3rd World Congress on Structural and Multidisciplinary Opimization, Buffalo, New York, USA, 1999.

[9] E. Kita, H. Tani, Shape optimization of continuum structures by genetic algorithm and boundary element method, *Engineering Analysis with Boundary Elements*, Vol. **19**, 1977.

[10] M. Kleiber (ed.) *Handbook of Computational Mechanics*, Springer, 1998.

[11] Z. Michalewicz, *Genetic Algorithms + Data Structures = Evolutionary Programs*. Springer-Verlag, AI Series, New York, 1992.

ADAPTIVE PENALTY STRATEGIES IN GENETIC SEARCH FOR PROBLEMS WITH INEQUALITY AND EQUALITY CONSTRAINTS

Chyi-Yeu Lin and Wen-Hong Wu
National Taiwan University of Science and Technology
43 Keelung Road, Section 4, Taipei, Taiwan 10672, Republic of China
jerrylin@mail.ntust.edu.tw

Abstract: This research aims to develop effective and robust self-organizing adaptive penalty strategies (SOAPS and SOAPS^e) for genetic algorithms to handle constrained optimization problems without the need of searching for proper values of penalty factors for a given optimization problem in hand. The proposed strategies are based on the idea that the constrained optimal design is almost always located at the boundary between feasible and infeasible domains. Both adaptive penalty strategies automatically adjust the value of the penalty parameter used for each of the constraints according to the ratio between the number of designs violating the specific constraint and the number of designs satisfying the constraint. The goal is to maintain equal numbers of design in each side of the constraint boundary so that the chance of locating their offspring designs around the boundary is maximized. Illustrative examples showed consistently improved performance on locating the global optimum in the problem with only inequality constraints and in the problem with both inequality and equality constraints.

Keywords: genetic algorithms, penalty function, constraint handling, constrained optimization, equality constraint, inequality constraint.

1.   INTRODUCTION

Most genetic algorithms were used as a tool of solving unconstrained optimization problems at the early stage of development [1], [2], [3]. When

241

T. Burczyński and A. Osyczka (eds),
IUTAM Symposium on Evolutionary Methods in Mechanics, 241–250.
© 2004 Kluwer Academic Publishers.

researchers started to apply genetic algorithms to design optimization problems of varied disciplines, it was the time that strategies had to be invented so that genetic algorithms could handle constraints encountered in most real-life design problems.

The big step of extending the capabilities of genetic algorithms to solve constrained problems in an efficient manner is the use of the penalty function, borrowed from the traditional sequential unconstrained minimization techniques (SUMT). In SUMT, a pseudo-objective function is defined by adding a penalty due to constraint violation on the original objective function. The value of the penalty depends on the degree of constraint violations, which is zero for a feasible design in the most often used exterior penalty function method. An infeasible design with a violation on a given constraint will cause an amount of penalty to be added on the objective function.

Many penalty function methods that required the users to define proper values for penalty parameters worked well in their illustrative problems but the performance may be degraded in other problems. Furthermore, in order to effectively apply to general optimization problems, the penalty function of a robust algorithm needs to be problem-independent. Most penalty strategies that have been proposed, unfortunately, are more or less problem-dependent. Michalewicz and Schoenauer [4] noted that although many sophisticated penalty function methods were reported as favorable and promising tools for constrained genetic searches, the traditional static penalty function method was found relatively reliable and robust due to simplicity of the penalty function. More detailed description of existing penalty function techniques for constrained genetic searches was provided [5], [6]. It is clear that due to the difficulties of proper definition of penalty parameters, there are currently two major trends of approaches that work on elevating the search effectiveness of genetic algorithms on constrained optimization problems. The first trend is the development of parameter-free, self-adaptive penalty function based methods. The second trend that avoids penalty functions uses special operators to create crossover pairs that contain members from the feasible domain and the infeasible domain.

In the rest of this paper, a parameter-free and problem-independent penalty function method, SOAPS, that seeks to maintain equal populations on either side of each constraint boundary is proposed. A special treatment is then applied on the SOAPS algorithm so as to improve search effectiveness on problems with equality constraints, and thus creating the SOAPS^e algorithm. The strengths of these strategies are demonstrated in two illustrative problems.

2. SELF-ORGANIZING ADAPTIVE PENALTY STRATEGY - SOAPS

Consider the following constrained minimization problem:

$$\text{Minimize: } F(\mathbf{x})/F' \tag{1}$$

$$\text{Subject to: } g_j(\mathbf{x}) = \frac{\overline{g_j(\mathbf{x})}}{b_j} \square 1 \square 0, \quad j = 1, m \tag{2}$$

$$x_i^L \square x_i \square x_i^U, \quad i = 1, n \tag{3}$$

where F^* is the normalization factor for the objective function.
The pseudo-objective function is defined as follows:

$$\text{Minimize: } \% (\mathbf{x}) = F(\mathbf{x}) + P(\mathbf{x}, q) \tag{4}$$

$$P(\mathbf{x}, q) = \frac{100 + q}{100} \cdot \frac{1}{m} \cdot \sum_j r_j^q \cdot \square g_j \tag{5}$$

where q is the generation number, r_j^q the penalty parameter for the jth constraint at generation q, and $\square g_j$ the constraint violation for the jth constraint. The iterative penalty parameters can be updated in the following manner:

$$r_j^q = r_j^{q\square 1} \cdot \left[1 \square \frac{(f_j^q \square 0.5)}{5} \right], \quad q \geq 1 \tag{6}$$

where f_i^q is the feasibility ratio for the jth constraint at generation q. At the first generation, $q=1$, the initial penalty parameter for the jth constraint is defined as follows:

$$r_j^0 = \frac{QR_{obj}^1}{QR_{con\,j}^1} \tag{7}$$

where QR_{obj}^1 is the interquartile-range of the objective function values of initial population, and $QR_{con\,j}^1$ interquartile-range of the jth constraint violation values of initial population.

The interquartile range of the objective function values of designs in the first generation is defined according to the following steps. Designs in the first generation are arranged into a sequence based solely on their objective function values. The design placed in the middle of the sequence is called the median design. The function value difference between the design placed one quarter of the population size ahead the median design and the design placed one quarter of the population size behind is defined as the interquartile range of the objective function values for the first generation. An illustrative example of the interquartile range of a function value of a population is shown in Fig.1. The interquartile range of the function of the population is 55, the difference between 23 and 78.

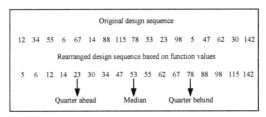

Figure 1. Illustration of interquartile range of a design sequence.

In a similar approach, the interquartile range for each of the constraints can be attained for the initial population. By using the ratio between quartile ranges of the constraint and the objective function, proper normalization are automatically achieved without a prior knowledge on the objective and constraint function distributions.

In each generation, the penalty parameter of each constraint is increased or decreased according to the feasibility percentage of each constraint. The term $(100+q)/100$ serves to increase the pressure on infeasible designs along evolution so as to gradually move the location of the minimum design for the pseudo-objective function back to where the true constrained optimum is. It is noted that no parameters in the penalty function need to be defined by users. Furthermore, the normalization between the objective and constraint functions is automatically achieved. It is a parameter free penalty function.

3. EQUALITY CONSTRAINT CONSIDERATION - SOAPS[E]

Most engineering optimizations comprise only inequality constraints. Most constraint handling techniques are developed mainly to tackle inequality constraints. Equality constraints received insufficient attention and are relatively difficult to handle. Equality constraints are always active constraints on which the global optimum must reside. It is therefore the goal that designs are maintained and equally distributed on both side of each of the equality constraints. In SOAPS, the design distributions on each inequality constraint are taken into account and the penalty parameter is accordingly adjusted to create a pressure to form a 50%-50% distribution on either side of the inequality constraint. It is noted that it is only a pressure, not a compulsory execution to obtain an even distribution. Designs that are not located on the equality constraint line are subject to penalties, and the designs on the infeasible part of another inequality constraint will receive double penalties compared to designs located on the feasible part of the inequality constraint. This will create an extra migration pressure to move designs toward the pseudo-feasible part (not considering equality constraints) of the design space. The imbalanced design distribution caused by equality constraints cannot be resolved by SOAPS. Therefore, the self-organizing adaptive penalty strategy with equality constraint consideration, SOAPS^e, is created.

Consider the following constrained minimization problem:

$$\text{Minimize: } F(\mathbf{x}) \tag{8}$$

$$\text{Subject to: } \overline{h}_j(\mathbf{x}) = 0, \quad j = 1, k \tag{9}$$

$$\overline{g}_j(\mathbf{x}) \square 0, \quad j = 1, m \tag{10}$$

$$x_i^L \square x_i \square x_i^U, \quad i = 1, n \tag{11}$$

These constraints can be rewritten in the following form:

$$\text{Subject to: } g_{(2j\square 1)}(\mathbf{x}) = \overline{h}_j(\mathbf{x}) \square 0, \quad j = 1, k \tag{12}$$

$$g_{2j}(\mathbf{x}) = \square \overline{h}_j(\mathbf{x}) \square 0, \quad j = 1, k \tag{13}$$

$$g_{(2k+j)}(\mathbf{x}) = \overline{g}_j(\mathbf{x}) \square 0, \quad j = 1, m \tag{14}$$

Each equality constraint is firstly split into two inequality constraints as shown in Equation (12) and (13). The value of each of these two constraint functions is reduced by an amount that initially depends on the initial equality function value distribution and will decrease along the number of generation in an exponential manner. These two modified constraint functions for equality constraints are as follows:

$$g_{(2j\square 1)}(\mathbf{x}) = \overline{h}_j(\mathbf{x}) \square MID_j^{0+} \cdot EXP(\frac{\square 10q}{totalGen\#}) \square 0, \quad j = 1, k \tag{15}$$

$$g_{2j}(\mathbf{x}) = \square \overline{h}_j(\mathbf{x}) \square MID_j^{0\square} \cdot EXP(\frac{\square 10q}{totalGen\#}) \square 0, \quad j = 1, k \tag{16}$$

where MID_j^{0+} and $MID_j^{0\square}$ represents the median value of all positive function values, and all negative values, respectively, for the jth equality constraint function in the initial population and *totalGen#* is the expected total generation number in a genetic search.

4.   ILLUSTRATIVE PROBLEMS

The SOAPS is firstly tested on one numerical problem with only inequality constraint functions. The results of SOAPS on the testing problem are compared with Lin and Hajela's strategy [7] and Joines and Houck's approach [8], in which three parameters, C, α, and β, are exclusively set to as 0.5, 2 and 2, respectively. The SOAPS^e algorithm is then tested in a problem with both inequality and equality constraints. In the first test problem, each genetic search with a specific penalty strategy and a unique set of parameters is executed ten times by using different random seeds.

The first numerical test problem is defined as follows [9]:

Minimize:

$$F(\mathbf{x}) = [x_1^2 + x_2^2 + x_1x_2 - 14x_1 - 16x_2 + (x_3 - 10)^2 + 4(x_4 - 5)^2$$
$$+ (x_5 - 3)^2 + 2(x_6 - 1)^2 + 5x_7^2 + 7(x_8 - 11)^2 \qquad (17)$$
$$+ 2(x_9 - 10)^2 + (x_{10} - 7)^2 + 100]/100$$

Subject to:

$$g_1(\mathbf{x}) = \frac{4x_1 + 5x_2 - 3x_7 + 9x_8}{105} - 1 \square 0 \qquad (18)$$

$$g_2(\mathbf{x}) = \frac{3(x_1 - 2)^2 + 4(x_2 - 3)^2 + 2x_3^2 - 7x_4}{120} - 1 \square 0 \qquad (19)$$

$$g_3(\mathbf{x}) = \frac{10x_1 - 8x_2 - 17x_7 + 2x_8}{10} \square 0 \qquad (20)$$

$$g_4(\mathbf{x}) = \frac{x_1^2 + 2(x_2 - 2)^2 - 2x_1x_2 + 14x_5 - 6x_6}{100} \square 0 \qquad (21)$$

$$g_5(\mathbf{x}) = \frac{8x_1 + 2x_2 + 5x_9 - 2x_{10}}{12} - 1 \square 0 \qquad (22)$$

$$g_6(\mathbf{x}) = \frac{5x_1^2 + 8x_2 + (x_3 - 6)^2 - 2x_4}{40} - 1 \square 0 \qquad (23)$$

$$g_7(\mathbf{x}) = \frac{3x_1 + 6x_2 + 12(x_9 - 8)^2 - 7x_{10}}{100} \square 0 \qquad (24)$$

$$g_8(\mathbf{x}) = \frac{0.5(x_1 - 8)^2 + 2(x_2 - 4)^2 + 3x_5^2 - x_6}{30} - 1 \square 0 \qquad (25)$$

$$\square 10 \square x_i \square 10 \text{ for } i = 1 \sim 10 \qquad (26)$$

For all strategies, the precision of each design variable was set to 0.001, the population size was set to 100 and probability of crossover and mutation was set to 0.8 and 0.001, respectively. The convergence histories of averaged best feasible designs in 10 searches at the end of each generation by five search approaches are shown in Fig.2.

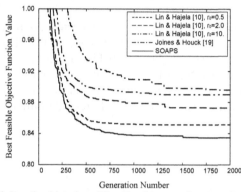

Figure 2. Iteration histories of five algorithms on the numerical problem.

Lin and Hajela's strategy was conducted with three different initial penalty factors: $r_0 = 0.5$, 2.0, and 10.0. Among them, smaller penalty factors produced better results as shown in Fig.2. The SOAPS resulted in the apparent best convergence history due to the unique capability of striving to maintain 50% feasible designs. Joines and Houck's approach had the worst search output in this problem.

The second test problem is defined as follows:

Minimize: $obj = x_1^{0.6} + x_2^{0.6} + x_3^{0.4} + 2x_4 + 5x_5 \square 4x_3 \square x_6 + 25$ (27)

Subject to:

$$h_1 = x_2 \square 3x_1 \square 3x_4 \qquad (28)$$

$$h_2 = x_3 \square 2x_2 \square 2x_5 \qquad (29)$$

$$h_3 = 4x_4 \square x_6 \qquad (30)$$

$$g_1 = x_1 + 2x_4 \square 4 \qquad (31)$$

$$g_2 = x_2 + x_5 \square 4 \qquad (32)$$

$$g_3 = x_3 + x_6 \Box 6 \tag{33}$$

$$x_1 \Box 3 \text{ , } x_3 \Box 4 \text{ , } x_5 \Box 2 \text{ , } x_1, x_2, x_3, x_4, x_5, x_6 \Box 0 \tag{34}$$

The problem has three equality constraints and three inequality constraints. The reported global solution is x^*=[0.67, 2, 4, 0, 0, 0] with an objective function of 13.04 [10]. A better global solution x^*=[0.167, 2, 4, 0, 0, 2] with $F(x^*)$=11.598 was found in the following genetic searches.

The results of the SOAPS^e algorithm in the second test problem are compared with Lin and Hajela's strategy with three different initial penalty factors: r_0 = 0.5, 2.0, and 10.0. and SOAPS. For all strategies, the population size was set to 200 and the probability of crossover and mutation was set to 0.6 and 0.001, respectively. Each genetic search with a specific penalty strategy and a unique set of parameters is executed 20 times by using different random seeds. The convergence histories of varied algorithms based on the average values are showed in Fig. 3. The number of global optimum reached, the number of feasible final designs, the best final function value, the average final function value and the worst final function value from twenty genetic searches by six different algorithms are reported in Table 1.

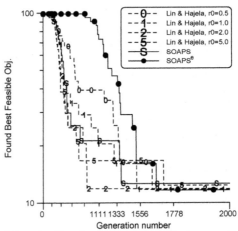

Figure 3. Iteration histories of six algorithms in the second problem.

Table 1. The statistical results on the second problem

METHOD	GLOBAL	FEA.	BEST	AVE	WORST
Lin and Hajela, $r^0=0.5$	7	15	11.598	11.843	12.218
Lin and Hajela, $r^0=1.0$	12	20	11.600	11.742	12.215
Lin and Hajela, $r^0=2.0$	9	20	11.606	11.845	12.484
Lin and Hajela, $r^0=5.0$	2	20	11.645	12.369	13.376
SOAPS	1	20	11.686	12.616	13.741
SOPAS^e	19	20	11.598	11.664	12.170

It is noted that from Fig. 3, the SOPAS^e has the slowest convergence history among all algorithms but this late convergence situation prolongs the search power to late stages of the evolution so that it can find the best final solution. Other algorithms often quickly crowd to sub-optimum solutions that leading to a fast decreasing average fitness function history, but it also prevents the algorithm from converging to the global optimum. From Table 1, it is noted that although SOAPS provides very effective capabilities in locating the global optimum on problems with only inequality constraints, it has a poor performance of finding the global optimum of the problem. Lin and Hajela's algorithms with varied penalty parameters have a combined less than 50% chance of locating the global optimum. The newly developed SOAPS^e produced an excellent record of locating the global optimum 19 times out of 20 trials. Furthermore, SOAPS^e also outperformed other algorithms in the best, the mean and the worst final designs from 20 executions.

5.　CONCLUSION

This paper proposed a self-organizing adaptive penalty strategy (SOAPS) for genetic algorithms to tackle constrained optimization problems. The parameter free characteristic of SOAPS further increase the performance efficiency since there is no need now to select any value for penalty parameters for a given problem in hand. SOAPS performed favorably in the illustrative problem with inequality constraints compared to other algorithms. Optimization problems with equality constraints are relatively difficult to solve, but they received insufficient consideration in the past. We also developed an effective adaptive penalty function strategy, SOAPS^e for problems with equality constraints. The strength of the equality specialist,

SOAPS^e, is demonstrated in another illustrative problem with equality constraints.

ACKNOWLEDGMENT

This research was supported by National Science Council of Republic of China, under grant, NSC89-2212-E-011-003.

References

[1] Holland, J. H., *Adaptation in Natural and Artificial System*, University of Michigan Press, 1975
[2] Goldberg, D. E., *Genetic Algorithms in Search, Optimization, and Machine Learning*, Addison-Wesley, 1989
[3] Gen, M. and Cheng, R., *Genetic Algorithms and Engineering Design*, John Wiley & Sons, 1996
[4] Michalewicz, Z. and Schoenauer, M., *Evolutionary Algorithms for Constrained Parameter Optimization Problems*, Evolutionary Computation, Vol.4, No.1, 1996, pp.1-32.
[5] Michalewicz, Z., *Genetic Algorithms, Numerical Optimization and Constraints*, Proceedings of the 6th International Conference on Genetic Algorithms, Morgan Kaufmann, 1995, pp.151-158
[6] Gen, M. and Cheng, R., *A Survey of Penalty Techniques in Genetic Algorithms*, Proceedings of IEEE International Conference on Evolutionary Computation, 1996, pp.804-809.
[7] Lin, C.-Y. and Hajela, P., *Genetic Algorithms in Optimization Problems with Discrete and Integer Design Variables*, Engineering Optimization, Vol.19, No.3, 1992, pp.309-27.
[8] Joines, J. and Houck, C., *On the Use of Non-stationary Penalty Functions to Solve Nonlinear Constrained Optimization Problems with GAs*, Proceedings of the First IEEE Conference on Evolutionary Computation, 1994, pp.579-84.
[9] Hock, W. and Schittkowski, K., *Test Examples for Nonlinear Programming Codes*, 187, Lecture Notes in Economics and Mathematical Systems, Springer-Verlag, 1981
[10] Floudas, C.A. and Pardalos, P.M., A collection of test problems for constrained global optimization algorithms, Springer-Verlag, 1990

ON THE IDENTIFICATION OF LINEAR ELASTIC MECHANICAL BEHAVIOUR OF ORTHOTROPIC MATERIALS USING EVOLUTIONARY ALGORITHMS

N. Magalhães Dourado, J. Cardoso Xavier, J. Lopes Morais
CETAV/UTAD, Departamento de Engenharias, Quinta de Prados, 5000-911 Vila Real, Portugal ndourado@utad.pt; jmcx@utad.pt; jmorais@utad.pt

Abstract: A numerical tool is presented to identify the linear elastic properties of orthotropic and homogeneous materials, coupled with the off-axis tensile test, as a part of a hybrid numerical-experimental model. The method combines whole-field displacement measurement techniques and an optimisation procedure based on the Finite Element Method and a developed Genetic Algorithm. A Finite Element analysis of the off-axis tensile test was performed using known in-plane linear elastic properties of wood pine *lodgepole* to generate a reference nodal displacement field to calibrate the numerical method. An objective function was chosen to minimise the mean quadratic difference between the reference displacement field and the displacement field that is calculated for each potential solution. A Sigma Truncation Scaling mechanism was chosen and a Genetic Algorithm with Varying Population Size (GAVaPS) was developed, based on an elitist strategy. Good approximation was acquired for the in-plane elastic properties.

Keywords: genetic algorithm with varying population size; heterogeneous fields; inverse method

1. INTRODUCTION

The in-plane elastic mechanical behaviour of orthotropic and homogeneous materials is characterized through four independent

T. Burczyński and A. Osyczka (eds),
IUTAM Symposium on Evolutionary Methods in Mechanics, 251–264.

engineering properties [1,2]: E_1, E_2, $\sqsubset_{12}$ and G_{12}, where the subscripts represent the material axes. Usually, two tensile tests, through directions 1 and 2, are performed to identify the Young *moduli* (E_1 and E_2) and the major *Poisson's* ratio ($\sqsubset_{12}$). Additionally, an in-plane shear test is carried out to obtain the shear modulus (G_{12}). Different shear test methods have been proposed to identify the shear *moduli* of an orthotropic material. Among them are the off-axis tensile test [3, 4], the Arcan test [5] and the Iosipescu test [6,7]. The use of whole-field displacement measurement techniques in combination with a suitable analytical or numerical tool, have been brought a new approach for the identification of the mechanical behaviour of orthotropic materials [8, 9, 10]. The aim of this approach is to reach a heterogeneous stress and strain fields through the specimen in a way that all elastic properties, that should play a balanced role in the response of the specimen, can be determined in a single test. In this work the off-axis tensile test was chosen to generate the heterogeneous stress and strain fields. A numerical identification procedure was developed to determine the in-plane elastic properties based on an Evolutionary Algorithm coupled with the Finite Element Method.

2.   INVERSE METHODS

Figure 1 illustrates the general inverse approach of identification of the linear-elastic mechanical behaviour of an orthotropic and homogeneous material, based on Evolutionary Search techniques and on the Finite Element Method. This approach is based on the experimental measurement of the specimen response and the application of a hybrid numeric-experimental method. Conceptually, this hybrid method consists in: a numerical model of the mechanical test, constructed using the Finite Element Method, and an iterative procedure used to search the elastic properties compatible with the measured displacement field.

Evolutionary Algorithms are nowadays an optimisation method widely used in inverse problems employing the above solution methodology [11, 6]. The inverse problem resolution enables to find out the set of project variables (*i.e.*, the material in-plane elastic properties) of the physical problem (*i.e.*, the off-axis mechanical test), which corresponds to the set of state known variables (*i.e.*, heterogeneous displacement field).

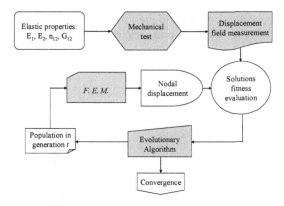

Figure 1. Identification of linear elastic behaviour of an orthotropic material using an inverse methodology

3.   FINITE ELEMENT METHOD ANALYSIS

The off-axis tensile test simulation by Finite Element Method was carried out for the test specimen geometry shown in Fig. 2, with $L=120$ *mm*, $w=30$ *mm*, $t=10$ *mm* and $\square=20°$, using the commercial code *ANSYS 6.0*[1]. A reference displacement field was firstly generated using the in-plane elastic properties of wood pine *lodgepole* [12]: $E_1=10,12\,GPa$, $E_2=1,032\,GPa$, $G_{12}=0,496\,GPa$ and $\square_{12}=0,316$. This displacement field was used to calibrate the numerical method, playing the role of the experimental data input, for the identification algorithm.

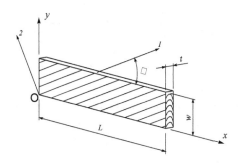

Figure 2. Off-axis test specimen geometric parameters

The boundary conditions applied to the Finite Element Model (see Fig.3) are in agreement with the rigid and non-rotating testing machine grips. Left-end nodes were fixed and a nodal displacement prescription $u_x = 0,5$ *mm* was applied to right-end nodes.

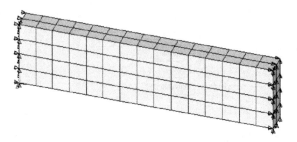

Figure 3. Finite element model and boundary conditions of the off-axis tensile test specimen

A typical *"S"* deformed shape was observed for the off-axis uniaxial tensile test specimen after *FEM* analysis (*Fig. 4a*). The obtained reference displacement field, u_x and u_y (*Fig. 4b*), exhibits a clear heterogeneity.

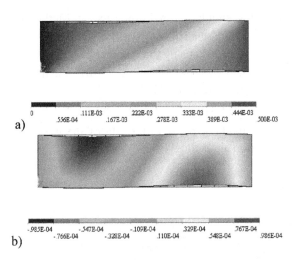

Figure 4. Reference displacement field (a) u_x and (b) u_y

4. GENETIC ALGORITHM

4.1 Introduction

Among all evolution-based search algorithms, the Genetic Algorithms (GAs) are perhaps the most well known. GAs were developed by John Holland [13,14] in an attempt to explain the adaptive processes of natural systems and to design artificial systems based upon these natural systems. Although not being the first algorithms to use principles of natural selection and genetics within the search process, GAs are today the most widely used [15]. More experimental and theoretical studies have been made on the field of GAs than on any other Evolutionary Algorithms (EAs).

During the last thirty years there has been a growing interest in problem resolution strategies of different kinds of problems based upon the principles of natural evolution and hereditary laws. These problem solving strategies favour a population of potential solutions,

The aim of Genetic Algorithms is to identify the individual (the set of elastic properties) best adapted to the environment (according to a fitness function), among the populations (set of potential solutions) found in successive generations (iterations).

One variety of these systems is a class of Evolution Strategies – algorithms that mimic the main principles of natural evolution for parameter optimisation problems [16]. Fogel´s Evolutionary Programming [17] is a technique for searching through a space of small finite-state machines. Glover´s scatter search techniques [18] hold a population of reference points and breed offspring based on weighted linear combinations. Holland´s GAs set up another type of evolution-based systems.

An EA is a probabilistic algorithm that sustains a population of solutions $P(t) = \left\{ x_1^t, \ldots, x_n^t \right\}$, in each generation t (see Fig. 5). Each individual x_i^t represents a potential solution to the problem and is encoded according to a predefined data structure S. A fitness value is determined according to how well each solution x_i^t fulfils objective function, y, of the problem. Then, a new generation is formed $(t = t + 1)$ selecting the set of more fit solutions (*selection* operator). Members of this emerging generation experienced transformations produced by genetic operators. These transformations may be arranged into two types: unary transformations m_i (*mutation* type) and higher order transformations c_j (*crossover* type) [19]. The former is characterized by a small change in single individual, introducing some extra variability into the population, and the later by new individuals generated combining parts from several (two or more) individuals $\left(c_j : S \cdots \cdots S3 \ S\right)$. Population size may remain constant throughout the algorithm or vary according to any birth-rate control strategy. After some number of generations the program converges. The best solution found till then is hoped to represent a near-optimum (reasonable) solution.

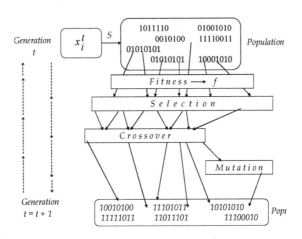

Figure 5. Illustration of a Genetic Algorithm scheme

4.2　　Developed Genetic Algorithm

As mentioned in section 3 a reference displacement field of off-axis tensile specimen (Reference solution in Fig. 6), $\mathbf{u}_i$, was firstly generated, using the in-plane elastic properties of wood pine *lodgepole*. Compliance matrix $[s]$ was previously determined, using the set of mentioned reference in-plane elastic properties, with $\square = 20°$. Solutions of the GA were represented by four-dimensional binary design variables ($E_1 , E_2 , G_{12} , \square_{12}$). Each variable x_i $(i=1,...,4)$ could take values from a predefined domain $D_i = [a_i , b_i] \square R$. The length ascribed to each variable m_i, based upon the precision p required to determine the objective function, y, was calculated (Bits number evaluation in Fig. 6) considering the smallest integer such that [19]:

$$\left(b_i \square a_i\right) \square 10^P \square 2^{m_i} \square 1 \tag{1}$$

An initial population with a number $PopSize(t)$, $(t=1)$, of potential solutions $P(1) = \{x_1^1, ..., x_k^1\}$, was randomly generated.

Each variable value (Decoding in Fig. 6) is given by the equation

$$x_i = a_i + decimal(string_2) \cdot \frac{b_i \square a_i}{2^{m_i} \square 1} \tag{2}$$

where $decimal(string_2)$ represents the hexa-decimal value of $(string_2)$. The objective function evaluation $y[i] = eval^t(v_i)$, of each chromosome v_i $(i = 1, ..., PopSize(t))$, was carried out using

$$y[i] = \frac{1}{N}\left\{ \sum_{i=1}^N \left\| \mathbf{u}_i \square \mathbf{u}_i^t \right\|^2 \right\} \tag{3}$$

where N represents the total number of nodes in the front plane of the off-axis test model (Fig. 3), $\mathbf{u}_i$ the reference displacement field and $\mathbf{u}_i^t$ the displacement field obtained by *ANSYS* for solution v_i, in generation t. Compliance matrix was up-dated for each $PopSize(t)$ solution.

4.2.1 Selection operator

Population ranking was then performed according to the values of the objective function, y, calculated using Eq. (3), and the $PopSize(t)$ solutions were structured in two subsets: S_T and S_R (Fig. 6). The quantities $n_T(t)$ and $n_R(t)$ represent the cardinal of subsets S_T and S_R, respectively, for generation t.

$$
\left.
\begin{array}{c}
(x_1^{t,1} \ldots\ldots x_k^{t,1}) \\
\cdots\cdots \\
\cdots\cdots \\
(x_1^{t,n} \ldots\ldots x_k^{t,n})
\end{array}
\right\} \; n = n_T(t) \; s_T
$$

$$
\left.
\begin{array}{c}
(x_1^{t,n+1} \ldots x_k^{t,n+1}) \\
\cdots\cdots\cdots \\
\cdots\cdots \\
\cdots \\
\cdot \\
(x_1^{t,PopSize(t)} \; x_k^{t,PopSize(t)})
\end{array}
\right\} \; n_R(t) \; s_R
$$

Figure 6. Population after Ranking

4.2.2   Scaling operator

A scaling mechanism (Scaling in Fig. 7) was used to improve the sensitivity of the algorithm to find solutions which exhibit objective function values very close to each other. Thus, a previous evaluation of the maturity state of convergence process was performed, and a Sigma Truncation scaling mechanism was chosen [19], according to

$$
y'[i] = y[i] + \left(AvgObj \; \square \; c \cdot \square \right) \tag{4}
$$

where c is chosen as a small integer, $\square$ is the population's standard deviation and *AvgObj* represent average objective function values, in the current population. Possible negative evaluations $y'[i]$ are set to zero.

4.2.3   Crossover operator

A probability of crossover p_c was previously assumed and an expected number $p_c \cdot PopSize(t)$ of chromosomes to undergo the crossover operation was then determined. An even number of chromosomes was assured by adding or subtracting one solution from this mapping pool – subset of

$p_c \cdot PopSize(t)$ solutions from the current population. The decision to add or subtract a solution from this mapping pool was done randomly. Thus, once taken the decision to add a previously rejected solution from the population, the chromosome selected to figure in the mapping pool was also done randomly.

Pairs of solutions were randomly settled from this group of chromosomes and crossing points were randomly assorted for each pair, from the range $[1..m\text{-}1]$, with $m = \sum_{i=1}^{k} m_i$.

Features' combination of different chromosomes to form two similar offspring solutions, each other, was performed swamping corresponding segments of the parents, *i.e.* , for a pair of m-dimensional vectors $\left(x_1^{t,a} \; x_2^{t,a} \ldots x_m^{t,a} \right)$ and $\left(x_1^{t,b} \; x_2^{t,b} \ldots x_m^{t,b} \right)$, crossing the chromosomes after the first gene produce the offspring $\left(x_1^{t,a} \; x_2^{t,b} \ldots x_m^{t,b} \right)$ and $\left(x_1^{t,b} \; x_2^{t,a} \ldots x_m^{t,a} \right)$.

A lifetime parameter for an i-th offspring solution, $(lifetime[i])$, was then determined by proportional allocation [19]:

$$minimum \left(MinLT + \beth \; \frac{AvgObj}{y[i]} \; , \; MaxLT \right) \qquad (5)$$

MinLT and *MaxLT* stand for maximal and minimal allowable lifetime values, respectively. *AvgObj* represent average objective function values, in the current population (Statistical parameters in Fig. 7), and

$$\beth = \frac{1}{2} \left(MaxLT \; \square \; MinLT \right) \qquad (6)$$

4.2.4   Mutation operator

Mutation operator was performed on a bit-by-bit basis and a probability of mutation p_m was assumed. There were a total number of $PopSize(t) \cdot m$ bits in the whole population and an expected (on average) number $PopSize(t) \cdot m \cdot p_m$ of mutations per generation. Every bit had an equal chance to be mutated. Thus, for every bit in the population $P(t)$, a random (float) number r was generated from the range $[0,....,1]$. Then, a bit mutation took place if $r < p_m$.

Objective function $y[i]$ and lifetime parameter $(lifetime[i])$ values were also recalculated (Eqs. 4 and 5) for recently born (mutated) solutions. A similar criterion was used to up-date the objective function values $y[i]$ of the current population as in 4.2.2, according to the algorithm maturity (Fig.7).

4.2.5   Aging and elimination operator

Lifetime parameter ($lifetime[i]$) was up-dated according to

$$lifetime\left[\,i\,\right]_t = lifetime\left[\,i\,\right]_{t\square 1}\ \square\ 1 \tag{7}$$

Solutions belonging to subset S_R (Fig. 6) are not up-dated, in terms of the lifetime parameter - elitist strategy.

Population size in generation $(t+1)$, is also up-dated, eliminating solutions $E(t)$ which reached null lifetime parameter values, and adding the offspring solutions generated in *Crossover* (4.2.3), $Offspring(t)$, according to

$$PopSize(t+1)\ = PopSize(t)\square E(t) + Offspring(t) \tag{8}$$

A full search is done, eliminating clone solutions among the current population.

Figure 7 resumes all the steps of the developed Evolutionary Algorithm, described above.

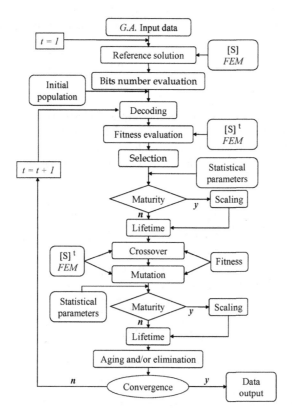

Figure 7. Developed Genetic Algorithm

5.   RESULTS AND DISCUSSION

In the present work an initial population of 10 individuals was used. The domain considered for each project variable was:

$$E_1 \square \left[9020,15070\right] \cdot 10^6 \left(Pa\right), \; E_1 \square R \; E_2 \square \left[867,1537\right] \cdot 10^6 \left(Pa\right), \; E_2 \square R$$

$$\square_{12} \square \left[280,392\right] \cdot 10^{\square 3}, \; \square_{12} \square R \; G_{12} \square \left[400,1351\right] \cdot 10^6 \left(Pa\right), \; G_{12} \square R.$$

which corresponds to the characteristic range of elastic properties of wood pine *lodgepole* [12].

The scaling operator was activated after 20 generations, for $c = 3$ in Eq. 4, and a total of 40 generations was achieved.

Lifetime parameter, $lifetime[i]$, (Eq. 5) was determined considering $MinLT = 1$ and $MaxLT = 7$.

A probability of crossover, $p_c = 0.25$, and a probability of mutation $p_m = 0.001$, were chosen.

Table 1 shows the set of values of the in-plane elastic property found for the best solution after 40 generations. The relative errors were determined with respect to the reference values presented in section 3. Three of them are inferior to the coefficient of variation associated to each of them, which is less than 22% [12].

Table 1. Elastic properties obtained according to the numerical method

Design variables	Elastic properties: (pine *lodgepole*)	Elastic properties (numerical method)	Relative error [%]
$E_1\ [GPa]$	10.120	10.335	2.13
$E_2\ [GPa]$	1.032	1.171	13.50
$\square_{12}$	0.316	0.320	1.36
$G_{12}\ [GPa]$	0.496	0.299	39.72

Fitness function, f, was defined as [20]:

$$f = k \;\square\; y \tag{9}$$

k being an arbitrarily large positive value that ensures that *fitness*, f, never becomes negative.

The value of the relative error found for the shear modulus, G_{12}, is greater than the typical scatter of experimental values of the elastic properties of wood. This large error can be attributed to the fact that the number of generations was not enough to ensure the stabilisation of the fitness value of the best solution (Fig. 8).

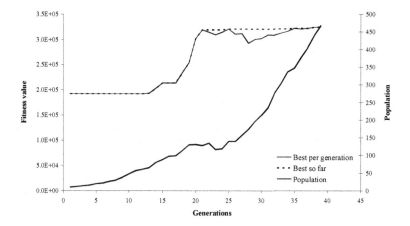

Figure 8. Fitness evolution of the best solution

Thus, the design variables shown in table 1 are not a suitable solution of the optimisation problem, from the point of view of the experimental scatter of mechanical properties of wood.

6. CONCLUSIONS

The numerical method developed to identify the elastic properties of orthotropic materials led to good approximation for the Young *moduli*: E_1 and E_2, and the Poisson's ratio, ν_{12}, determination. A great relative error was found for the shear modulus, G_{12}.

ACKNOWLEDGEMENTS

We would like to thank the Portuguese Foundation for Science and Technology for the financial support necessary to the execution of this work in the ambit of the project POCTI/1999/EME/36270.

References

[1] Tsai S. and H. Hahn, "Introduction to composite materials", Technologic Publishing C.O., 1980.
[2] Isaac, M. and I. Ori, "Engineering mechanics of composite materials", Oxford University Press, 1994.

[3] Kawai, M., *et al.*, "Effects of end-tab shape on strain field of unidirectional carbon/epoxy composite specimens subjected to off-axis tension", Composites Part A, 1997, 28A, pp. 267-275.

[4] Sun, C.T. and S. P. Berret, "A new end tab design for off-axis tension test of composite materials", Journal of Composite Materials, August 1998, 22, pp. 766-779.

[5] Hung, S.-C. and K. M. Liechi, "An evaluation of the Arcan Specimen for determining the shear *moduli* of fiber-reinforced composites", Experimental Mechanics, 37 (4), 1997, pp. 460-468.

[6] Liu, G. R. and Chen S. C., "Flaw detection in sandwich plates based on time-harmonic response using genetic algorithm", Computer Methods in Applied Mechanics and Engineering, 190, 5505 – 5514, 2001.

[7] Yoshihara, H, *et al.*, "Comparisons of shear stress/shear strain relations of wood obtained by Iosipescu and torsion tests", Wood and Fiber Science, 33(2), 2001, pp.275-283.

[8] Grédiac M. and Pierron F., "A T-shaped specimen for the direct characterization of orthotropic materials", International Journal for Numerical Methods in Engineering, Vol. 41, 239-309, 1998.

[9] Mota Soares C., Moreira de Freitas, M, Araújo and A. L. Pederson, "Identification of materials properties of composite plate specimen", Composite Structures, 25, 227-285, 1993.

[10] Le Magorou, F. Bos and F. Rouger, "Identification of constitutive laws for wood-based panels by means of an inverse method", Composites Science and Technology, vol. 62, n° 4, pp. 591-596, 2002.

[11] He Y., Guo D. and Chu F., "Using genetic algorithms to detect and configure shaft crack for rotor-bearing system", Computer Methods in Applied Mechanics and Engineering, 190, 5895 – 5906, 2001.

[12] Forest Products Laboratory, "Wood handbook – Wood as an Engineering material" Gen. Tech. Rep. FPL – GTR – 113, U.S. Department of Agriculture, 1999.

[13] Holland, J.H., "Genetic Algorithms and the Optimal Allocations of Trials", SIAM Journal of Computing 2:2, 88-105, 1973.

[14] Holland, J.H., "Adaptation in Natural and Artificial Systems", University of Michigan Press, Ann Arbor, 1975.

[15] Dawkins, R., "Evolutionary Design by Computers", Ed. Peter J. Bentley, ISBN: 1-55860-605-X, 1999.

[16] Schwefel, H. –P., "Numerical Optimization for Computer Models", John Wiley, Chichester, U.K., 1981.

[17] Fogel, L. J., Owens, A. J. and Walsh, M. J., "Artificial Intelligence Through Simulated Evolution", John Wiley, Chichester, U.K., 1966.

[18] Glover, F., "Heuristics for Integer Programming Using Surrogate Constraints", Decision Sciences, Vol. 8, No. 1, pp. 156-166, 1977.

[19] Michalewicz, Z., "Genetic Algorithms + Data Structures = Evolution Programs", Springer-Verlag, 1999, pp. 33-80.

[20] Jenkins, W. M., "An enhanced genetic algorithm for structural design optimization", Neural Networks and Combinatorial Optimization in Civil and Structural Engineering edited by B. Topping and A. Khan, 1993, pp. 109-126.

RANKING PARETO OPTIMAL SOLUTIONS IN GENETIC ALGORITHM BY USING THE UNDIFFERENTIATION INTERVAL METHOD

Jerzy Montusiewicz

Department of Technology Fundamentals, Technical University of Lublin, 20-618 Lublin, 38a Nadbystrzycka St., POLAND. Email: montus@antenor.pol.lublin.pl

Abstract: The article shows a new method of ranking Pareto optimal solutions, which form a numerous set of nondominated solutions, by using the notion of optimality in the sense of an undifferentiation interval. The ranking algorithm presented is based on the filtration of a set of Pareto optimal solutions by using the undifferentiation interval method. The example presented shows that the generated subsets of nondominated solutions are given different ranks, which should contribute to an adequate crossover operation.

Keywords: genetic algorithm, multicriterial optimization, undifferentiation interval method, ranking method

1. INTRODUCTION

Genetic algorithms combine the evolutionary principle of the survival of the best-adapted artificial organisms with the principle of a systematic exchange of information. In every basic iteration of genetic algorithms there occurs a selection, crossover and mutation of solutions, which results in generating a new set of solutions composed of the combined fragments of the best-adapted representatives of the previous generation. Reference [5, 6] shows an application of the undifferentiation interval method in the selection of Pareto optimal solutions. The present article uses the method in ranking Pareto optimal solutions for the purpose of their adequate crossover. The basis of many methods of multicriterial optimization is the definition of an optimum in Pareto sense.

265

T. Burczyński and A. Osyczka (eds),
IUTAM Symposium on Evolutionary Methods in Mechanics, 265–276.
© *2004 Kluwer Academic Publishers.*

1.1   Pareto optimality

A solution is Pareto optimal if none of the criteria $F_1(\mathbf{x})$, $F_2(\mathbf{x})$, $F_m(\mathbf{x})$ can be improved on without a simultaneous deterioration of at least one of them. An element $\mathbf{x}^* \square X$ is called Pareto optimal if and only if there exists no $\mathbf{x}^- \square X$ such that for every $m \square 4$, $F_m(\mathbf{x}^-) \square F_m(\mathbf{x}^*)$, and there exists a $j \square 4$, such that: $F_j(\mathbf{x}^-) < F_j(\mathbf{x}^*)$.

1.2   Goldberg's ranking method

The method consists in an interactive selection of Pareto optimal solutions from the population subset Q and ascribing to them the same rank [1].

In step 1 the whole population Q is considered — Pareto optimum solutions (population Q_{P1}) receive rank 1.

In step 2 the rest of population $Q_{P2} = Q - Q_{P1}$ is considered, out of which Pareto optimum solutions are selected which are given rank 2.

In subsequent steps $(3, 4, ..., j)$ the remaining part of the population is always considered:

$$Q_{P3} = Q \square \sum_{i=1}^{2} Q_{Pi} \ , \ Q_{P4} = Q \square \sum_{i=1}^{3} Q_{Pi} \ , \ (...), \ Q_{Pj} = Q \square \sum_{i=1}^{j\square 1} Q_{Pi} \ .$$

The Pareto optimal solutions determined are given the respective ranks of 3, 4, ..., j. The process is continued until all the solutions in a given population Q are exhausted. The value of the fitness function f_G is calculated by means of the formula:

$$f_G(\mathbf{x}) = -r(\mathbf{x}) + j + 1 \tag{1}$$

where: $r(\mathbf{x})$ — rank given to solution $\mathbf{x}$,
          j — number of generated subset of different ranks.

Fitness function $f_G(\mathbf{x})$, expressed by relation (1) always assumes positive values. Maximization of the value of the function prefers individuals of minimal rank value $r(\mathbf{x})$ (the lower the rank value, the higher the value of the $f_G(\mathbf{x})$ function). According to its rank, it is possible to assign to each solution a probability of reproduction or a number of descendant copies.

2.   NEW RANKING METHOD

The proposed ranking method consists in selecting Pareto optimal solutions. The selection process always requires particular attention since it is a basic operation in genetic algorithms and it determines the proper functioning of the algorithm.

A properly performed selection will ensure that the crossover operation will bring the desired results. The literature [1, 2, 3, 4, 7, 8, 9, 10, 11] describes many selection methods and their application in various optimization tasks. It should be remembered that the selection process in genetic algorithms is connected with two correlated questions: selection pressure and population diversity. Strong selection pressure, or intensive promotion of the best generations, reduces population diversity, which contributes to the loss of precious information about the remaining potential solutions and leads to premature convergence. Weak selection pressure, or great population diversity, results in an ineffective search.

The method proposed is addressed to situations where the set of Pareto optimal solutions is very large and has many solutions, e.g. several tens. With a larger number of solutions we should strive to reduce the set's size by applying appropriate filters, e.g. [5, 6].

The method consists in interactive determination of subsets of nondominated solutions in the undifferentiation interval sense out of a population subset P and giving them the same rank.

2.1 Optimality in undifferentiation interval sense

The notion of optimality applied here is that of an undifferentiation interval (UI), expressed as a percentage of the value of the criterion analyzed [5, 6]. The multicriterial analysis is carried out in criterial space and aims at determining whether a Pareto optimal solution 'deteriorated' by the accepted undifferentiation interval remains a nondominated solution and is added to the currently created subset of nondominated solutions. An element $\mathbf{x}^{\wedge} \square \square$ will be a nondominated solution in the undifferentiation interval sense if and only if $\square$ does not contain an element $\mathbf{x}^{+}$ such that there exists an $l \square 4$ for which:

$$\text{when:} \quad F_1(\mathbf{x}^{\wedge}) \square 0: \quad F_1(\mathbf{x}^{\wedge}) < F_1(\mathbf{x}^{+}) \text{ and } \left(1 + \frac{UI}{100}\right) F_1(\mathbf{x}^{\wedge}) > F_1(\mathbf{x}^{+})$$

$$(2)$$

$$\text{when:} \quad F_1(\mathbf{x}^{\wedge}) < 0: \quad F_1(\mathbf{x}^{\wedge}) < F_1(\mathbf{x}^{+}) \text{ and } \left(1 - \frac{UI}{100}\right) F_1(\mathbf{x}^{\wedge}) > F_1(\mathbf{x}^{+})$$

where: $\square$ — a non-empty set of Pareto optimum solutions.

Figure 1 presents a graphic illustration of how the undifferentiation interval works. In Figure 1a condition (2) is fulfilled and solution $\mathbf{x}^{\wedge}$ is rejected; in Figure 1b condition (2) is not fulfilled and solution $\mathbf{x}^{\wedge}$ remains as a nondominated solution in the undifferentiation interval sense (NSUIS).

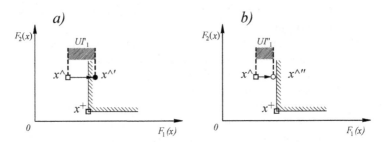

Figure 1. Graphical illustration 'deteriorating' the compared solution

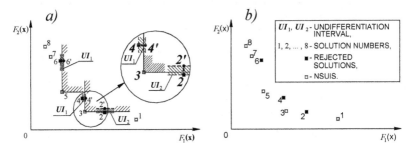

Figure 2. Determining nondominated solutions in UI sense with bicriterial minimization

Figure 2 presents the idea of determining the solutions nondominated in the undifferentiation interval sense described by condition (2) from an 8-element Pareto optimal solution set, with both criteria minimized. Figure 2a presents the function of the UI_1 for criterion $F_1(\mathbf{x})$, in relation to solutions 4 and 6, as well as UI_2 for criterion $F_2(\mathbf{x})$, in relation to solution 2. Let us consider in detail solutions 3 and 4. Solution 4 had a greater value of criterion $F_2(\mathbf{x})$ and a smaller value of criterion $F_1(\mathbf{x})$ than solution 3, and that was why it is nondominated in Pareto sense. Applying the undifferentiation interval method we temporarily 'deteriorate' the value of the $F_1(\mathbf{x})$ component of solution 4 by the UI_1, moving solution 4 along the F_1 direction to position 4'. In this situation, solution 3 dominates over the resulting solution 4'. The conclusion is that solution 4 is not nondominated in the undifferentiation interval sense and is therefore eliminated. The result of comparing solution 2 with solution 3 is the elimination of solution 2. In this case UI_2, referring to criterion $F_2(\mathbf{x})$, was decisive. Solution 2 moved along direction F_2 to position 2' which was dominated by solution 3. According to the same procedure, solution 5 eliminated solution 6. Figure 2b presents the subset of nondominated solutions in the undifferentiation interval sense. It is to be stressed that solutions 2 and 4 (for the accepted values of UI_1 and UI_2) cannot in any way eliminate solution 3.

2.1.1 Mutual exclusion phenomenon (MEP)

The mutual exclusion phenomenon occurs when both the solutions compared eliminate each other in the analysis of particular criteria. Solution **A** excludes solution **B** (Figure 3a) through the UI_1 related to criterion $F_1(\mathbf{x})$, while solution **B** excludes solution **A** (Figure 3b) through the UI_2 related to criterion $F_2(\mathbf{x})$. This means that they belong respectively to positive cones with the vertex in **A** (Figure 3a) and **B** (Figure 3b). In that case both the solutions compared are added to the created subset of solutions nondominated in the undifferentiation interval sense. The mutual exclusion phenomenon occurs first of all with relatively large values of the undifferentiation interval.

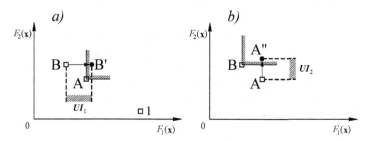

Figure 3. Mutual exclusion phenomenon

2.1.2 Undifferentiation interval correction (UIC)

The occurrence of the mutual exclusion phenomenon was eliminated by the application of a correction of the undifferentiation interval, consisting in an additional analysis of the same solutions – verification of condition (2) with the reduced value of the undifferentiation interval. The reason was the assumption that if lower values of the undifferentiation interval contribute to eliminating the compared solutions, it is an important property of those solutions and cannot be lost in the filtering process. Therefore we examine whether with lower values of the undifferentiation interval any of those solutions might be nondominated in the undifferentiation interval sense. If so it is added to the subset of solutions nondominated in the undifferentiation interval sense, and the other solution is rejected.

2.2 The method's algorithm

With a small differentiation of solutions in a population (Figure 4a) or their regular distribution (Figure 4b) it may happen that the crossover operation will not cause the displacement of new populations towards the ideal point of the set of evaluations of the multicriterial optimization task.

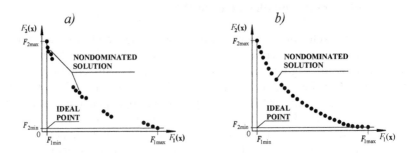

Figure 4. Graphical illustration of the Pareto optimal solution set: a) of a small differentiation
of solutions with an irregular distribution; b) regular distribution

The research carried out indicates that too often during the multicriterial optimization of the crossover solutions occur close to one another, which leads to a weak differentiation of the genetic material. The algorithm presented below is aimed at different ranking of consecutive solutions. Since the most frequent participants will be solutions with the highest value of the fitness function $f_G(\mathbf{x})$ (with the lower rank value of $r(\mathbf{x})$), they will not be consecutive solutions.

The ranking method presented is based on the notion of optimality in the sense of an undifferentiation interval and Goldberg's method. Shown below is the method's algorithm:

Step 1 Calculate the original undifferentiation interval, $j=1$, $UI_j = \dfrac{100}{P}$,

expressed in %, P — numerical force of the Pareto optimum set;

Step 2 Determine a subset of nondominated solutions in the undifferentiation interval sense with a given UI_j and give its elements a rank $r(\mathbf{x})=j$, P_j — the numerical force of the j-th subset of nondominated solutions in the undifferentiation interval sense;

Step 3 Check the STOP criteria:

   a) if $j = 1$, then go to step 4;

   b) if $j > 1$ and $\dfrac{P}{j} > P \square \left(\sum\limits_{k=1}^{j} P_k \right)$, then the remaining solutions of the

   Pareto optimal subset get the rank $j+1$, go to step 7;

   c) if $P_{j+1} = P \square \left(\sum\limits_{k=1}^{j} P_k \right)$, go to step 7;

Step 4 Calculate the new value of the undifferentiation interval, $j = j + 1$,

$$UI_j = \frac{UI_{j\square}}{2};$$

Step 5 Eliminate from the actual subset of Pareto optimal solutions those solutions ranked *j-1*;

Step 6 Go to step 2;

Step 7 STOP.

Attention should be drawn to the fact that the proposed selection of the existing Pareto nondominated solutions (according to the ranking method) is triggered at the moment when their number is sufficiently high (several tens of solutions). Ranking the existing Pareto nondominated solutions (i.e. their selection) concerns solutions only from the same population. Every inclusion of the ranking algorithm of particular Pareto nondominated solutions requires the calculation of the initial value of the undifferentiation interval UI_1 on the basis of the current size of the Pareto set.

3.   CALCULATION EXAMPLE

The calculation example has been taken from Osyczka [8] and concerns the multicriterial optimization of a multi-part beam (Figure 5).

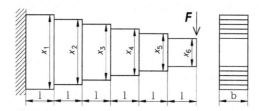

Figure 5. Scheme of the beam

The multicriterial problem was formulated as follows:

The vector of decision variables is: $\mathbf{x} = [x_1, x_2, \dots, x_6]^T$
  where: x_n is the thickness of *n*-th part of the beam.

The objective functions are:
  $F_1(\mathbf{x})$ – the volume of the beam [cm^3];
  $F_2(\mathbf{x})$ – displacement under force $\mathbf{F}$ [mm].

The constraints concerned permissible bending stresses in particular beam segments (six inequality constrains) as well as the thickness of particular segments (also six inequality constrains).

The problem was run for the following data:

  $l = 50$ [mm], $b = 50$ [mm], $\mathbf{F} = 10000$ [N], $E = 2.06 \times 10^5$ [N/mm^2],
  $s_g = 360$ [N/mm^2], $d = 32$ [mm].

The problem was considered as a discrete programming problem with the following set of values of decision variables:

$x_i = \{12,14,16,18,20,22,24,26,28,30,32\}$ [mm]   for $i=1,2,...,6$

The process of the multicriterial optimization of the above example was carried out by Osyczka by means of a genetic algorithm employing tournament selection.

3.1   Example 1

Indirect results of the calculation process served to test the proposed algorithm of ranking Pareto nondominated solutions. The method of ranking by using optimality in the undifferentiation interval sense was applied at a selected stage of the optimization process when the number of Pareto nondominated solutions was 66. In order to test the correctness of the presented algorithm of ranking Pareto optimal solutions and compare the obtained results, the calculations were repeated several times. The elements altered in the calculations were the initial value of the accepted undifferentiation interval and the algorithm determined the solutions optimal in the undifferentiation interval sense (algorithm testing condition (2)).

At the beginning the calculations were made for two values of the original undifferentiation interval:

☐ UI_1=1,5% (in accordance with step 1 – after rounding it off);
☐ UI_1=3,0% (twice as large as the result of the calculations specified in step 1) and for a value twice as large.

Table 1 shows the results obtained by using the algorithm which considered the mutual exclusion phenomenon, but had no built-in procedure for eliminating the consequences of the phenomenon. Thus the determined subset of solutions optimal in the undifferentiation interval sense was entered by mutually exclusive solutions, which resulted in the fact that contiguous solutions were ranked in the same way.

Table 1. Numerical force of subsets and their rank with MEP considered

	Original UI_1=1,5%		Original UI_1=3,0%	
	UI [%]	Number of elements	UI [%]	Number of elements
Rank 1	1,5	6	3	6
Rank 2	0,75	4	1,5	3
Rank 3	0,375	8	0,75	4
Rank 4	0,1875	13	0,375	6
Rank 5	0,09375	20	0,1875	12
Rank 6	0,046875	15	0,09375	20
Rank 7	STOP- Step 3 c)		0,046875	6
Rank 8			STOP- Step 3 b)	9

This fact is only visible in the figure showing the distribution of the ana-lyzed solutions in criterial space. Such a situation is particularly unacceptable for solutions receiving small ranking values (cf. Figure 6, detail A), in other words, large values of the fitness function $f_G(\mathbf{x})$), since these solutions will take part in the crossover process more often. The co-occurrence of highly-ranked solutions (cf. Figure 6, detail B), in other words, a small value of the fitness function $f_G(\mathbf{x})$, is not so important since those solutions will seldom take part in the crossover operation.

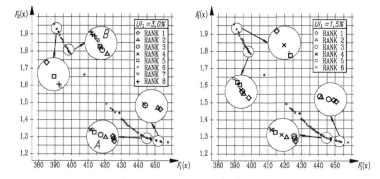

Figure 6. Graphical presentation of ranking results for UI_1=3,0% and UI_1=1,5% with MEP considered

Table 2 and Figure 7 show the results obtained by using the algorithm which took into account the mutual exclusion phenomenon and had the built-in procedure for eliminating the results of the phenomenon, i.e. the undifferentiation interval correction. When using this algorithm there is no situation of solutions with low ranking values occurring next to one another. Therefore the crossover operation should concern solutions lying far apart.

Table 2. Numerical force of subsets and their rank with MEP considered and UI correction

	Original UI_1=1,5%		Original UI_1=3,0%	
	UI [%]	Number of elements	UI [%]	Number of elements
Rank 1	1,5	5	3	3
Rank 2	0,75	3	1,5	3
Rank 3	0,375	9	0,75	3
Rank 4	0,1875	21	0,375	9
Rank 5	0,09375	21	0,1875	20
Rank 6	STOP- Step 3 b)	7	0,09375	21
Rank 7			STOP- Step 3 b)	7

Figure 8 shows the distribution of solutions which received the rank value of 1 and 2. It can be seen that those solutions cover the whole set of Pareto nondominated solutions and faithfully represent the shape of the set. Therefore the solution crossover operation should not lead to a loss of information about the whole set of Pareto optimal solutions.

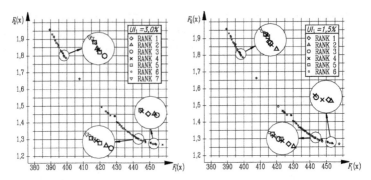

Figure 7. Graphical presentation of ranking results for UI_1=3,0% and UI_1=1,5% with MEP considered and UI correction

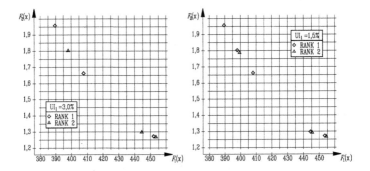

Figure 8. Graphical presentation of solutions ranking 1 and 2 with UI_1=3,0% and UI_1=1,5%, with MEP considered and UI correction

3.2   Example 2

The other case of calculation concerned a situation where the set of Pareto optimal solutions of the designed object had 40 elements. According to step 1 the initial value of the undifferentiation interval was 2,5% (UI_1=100/40=2,5).

Table 3. Numerical force of subsets and their rank with MEP considered and UI correction

	Original UI_1=2,5%			
	MEP		MEP with correction UI	
	UI [%]	Number of elements	UI [%]	Number of elements
Rank 1	2,5	6	2,5	3
Rank 2	1,25	4	1,25	4
Rank 3	0,625	3	0,625	6
Rank 4	0,3125	8	0,3125	11
Rank 5	0,15625	14	0,15625	13
Rank 6	STOP- Step 3 b)	5	STOP- Step 3 b)	3

The ranking procedure was repeated twice, using the algorithm considering mutually exclusive solutions and the algorithm with a built-in correction of the value of the undifferentiation interval. The results are contained in Table 3 and in Figure 9.

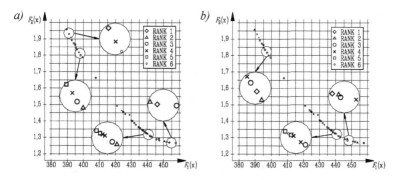

Figure 9. Graphical presentation of ranking results for the initial value of UI_1=2,5%: a) with MEP considered; b) with MEP considered and UI correction

4.   CONCLUSIONS

1. The method of ranking solutions from the Pareto optimal solution set, using the notion of optimality in the sense of an undifferentiation interval, allows a suitable selection of a large set of nondominated solutions.

2. Testing the correctness of the ranking algorithm presented was carried out by performing calculations for different initial values of the undifferentiation interval UI_1 and for two different sizes of the analyzed set of Pareto nondominated solutions: 66 elements and 40 elements (Tables 1, 2 and 3). Each time the algorithm carried out an effective selection of the set analyzed, assigning to every element an appropriate rank value.

3. The results presented in Table 1 and 2 show that introducing the initial value of the undifferentiation interval UI_1 in accordance with step 1 of algorithm leads to a quicker conclusion of the calculations than in the case of introducing a UI whose value is twice as large. In the former case the selection of the set analyzed leads to giving its elements a lower number of ranks.

4. A detailed analysis of Figures 6, 7 and 9, illustrating the distribution of elements with rank values assigned to them, leads to the conclusion that the selection operation should employ an algorithm with a built-in mechanism for correcting the value of the undifferentiation interval in order to eliminate solutions of the mutual exclusion phenomenon type.

5. From an analysis of Figure 8 it can be seen that solutions receiving the same rank value cover the whole set of Pareto nondominated solutions. Hence the crossover operation of those solutions will not lead to a loss of information about the whole Pareto set possessed.

References

[1] Goldberg D.E.: *Genetic Algorithms in Search, Optimization, and Machine Learning*, Addison-Wesley Publishing Company, Inc, 1989.

[2] Hancock P.: An Empirical Comparison of Selection Methods in Evolutionary Algorithms. In: *Evolutionary Computing*, Fogel D. (Ed.), Springer-Verlag, pp. 80-95, 1994.

[3] Kundu S. and Osyczka A.: The Effect of Genetic Algorithm Selection Mechanism on Multicriteria Optimization Using the Distance Method. In: *Theoretical and Applied Mechanics (ICTAM)* Kyoto, Japan, IUTAM Volume of Abstracts 272, 1996.

[4] Michalewicz Z.: *Genetic Algorithms+Data Structures=Evolution Programs*, Springer-Verlag Berlin Heidelberg, 1994.

[5] Montusiewicz J.: Division of the Set of Nondominated Solutions by Means of the Undifferentiation Interval Method, In: *The Technological Information Systems*, Societas Scientiarum Lublinensis, Lublin, pp. 65-72, 1999.

[6] Montusiewicz J.: Reducing the Pareto Optimal Set by Means of the Undifferentiation Interval Method, *Proceedings of the Second World Congress of Structural and Multidisciplinary Optimization*, vol.1, 26-30 May, Zakopane, pp. 97-102, 1997.

[7] Osyczka A., Krenich S. and Kundu S.: Proportional and Tournament Selections for Constrained Optimization Problems Using Genetic Algorithms, *Evolutionary Optimization an International Journal on the Internet*, 1 (1), pp. 89-92, 1999.

[8] Osyczka A.: Evolutionary Algorithms for Single and Multicri-teria Design Optimization, Physica-Verlag Heidelberg, 2002.

[9] Parmee I.C.: A Review of Evolutionary/ Adaptive Search in Engineering Design. Evolutionary Optimization an International Journal on the Internet, 1 (1), pp. 13-39, 1999.

[10] Schwefel H.P.: Evolution and Optimum Seeking, John Wiley and Sons, New York, 1995.

[11] Zhang B-T. And Kim J-J.: Comparison of Selection Methods for Evolutionary Optimization, *Evolutionary Optimization an International Journal on the Internet*, (2) 1, pp.54-69, 2000.

THE EFFECTIVENESS OF PROBABILISTIC ALGORITHMS IN SHAPE AND TOPOLOGY DISCRETE OPTIMISATION OF 2-D COMPOSITE STRUCTURES

A. Muc

Institute of Mechanics & Machine Design, Cracow University of Technology, Kraków, Poland

Abstract: The aim of the paper is to discuss the formulation and solution of optimisation problems for composite structures. A special attention is focused on the coding problems of design variables. The appropriate discrete coding allows us to use the same optimisation algorithms for different class of problems, i.e. shape and topology optimisation using both deterministic and fuzzy approaches.

Keywords: probabilistic algorithms, plates, shells, composites, fuzzy optimisation.

1. INTRODUCTION

Recently, an increasing trend to the application of different variants of evolutionary algorithms in searching for the optimal solutions in various classes of optimisation problems is observed. It is mainly caused by the simplicity of the proposed algorithm and on the other hand by the certainty that the found solution is better than the initial one.

In order to use anisotropic properties of composite materials in an appropriate manner a lot of efforts have been put into introduction and application of effective optimisation algorithms in the design of 2-D laminated plated and shell structures, particularly in the sense of searching for a global optimum. The latter problem is also associated with the imprecise definition of geometrical and material properties of composite materials what is taken into account via membership functions and so-called optimisation in a fuzzy enviroment. The second still open problem in this

T. Burczyński and A. Osyczka (eds),
IUTAM Symposium on Evolutionary Methods in Mechanics, 277–286.
© *2004 Kluwer Academic Publishers.*

area is associated with the effective definition of discrete design variables describing real continuous ones. The above-mentioned problems are discussed herein and various numerical examples are demonstrated. They are connected with the topology and shape optimisation.

The effectiveness of probabilistic algorithms is evaluated and determined in the sense of following factors:
– the accuracy of computed global optimum,
– the computational time required in the evaluation of the global optimum,
– the similarity (identity) of optimal design for various probabilistic algorithms.

Since for composite structures, especially in the discrete optimisation problems, a lot of local extrema exists, it is necessary to compare results with the use of various algorithms and in addition using different selection procedures. Both for shape and topology optimisation problems the optimisation procedures (probabilistic algorithms) are conjugated with the finite element analysis that is used for the numerical derivation of a required objective function – see Refs [1], [2]. However, they are completely separated from the FE analysis. In order to verify the effectiveness of the proposed algorithms three different approaches have been applied herein, i.e.: different variants of genetic algorithms (GA), simulated annealing (S.A.) algorithm and the variant of GA proposed in the EVOLVER package.

2.   OBJECTIVES OF OPTIMISATION

If any of design variables, constraints or even objective functions are not known precisely (uniquely – a crisp form) the formulation of the optimisation problem should be changed. It may be conducted using statistical analysis (not discussed herein) or the fuzzy formulation. In Table 1 there are compared two optimisation problems deterministic and fuzzy. Of course, the minimum design problem may be simply reformulated to the maximum problem. As it may be noticed the definition of the constraints is the fundamental difference in both formulations. However, as it will be demonstrated below the significant difference lies also in the optimisation algorithms and the reformulation of the fuzzy optimum design problems, since in a fuzzy environment searching for the optimal solutions is carried out not for the values of design variables s (a crisp problem) but for the membership functions μ.

Table 1. Comparison of the deterministic and fuzzy formulation of optimization problems

	Deterministic formulation	**Fuzzy formulation**
Objective	Find: Min f(s)	Find: Min f(s)
Constraints	$g_j(s) \le b_j$, $j = 1,2,...,m$	$g_j(s) \ \square \ G_j$ (fuzzy sets), $j=1,2,...m$

Thus, using the fuzzy formulation the initial minimum problem is formulated as a MinMax problem. The latter problem is reformulated to the following maximum problem introducing an additional objective $\square$:

$$\text{Find Max} \square \tag{1}$$

subject to the constraints

$$\lambda \le \mu_f(s), \quad \lambda \le \mu_{gj}(s), j=1,2,...,m. \tag{2}$$

3.   CODING OF THE DESIGN VARIABLES

Dependly on the analysed optimisation problem (Table 3) the binary coding of design variables is used with the aid of the incidence matrix introduced in the form given in Table 2. The total number of columns represents the length of the chromosome. Each gen in the chromosome describes the coded information in rows, i.e. if $a(1,1)=1$ then the ply orientation is equal to 0^0, and if $a(2,1)=a(3,1)=0$ then the ply orientation is not equal to 45^0 (or 90^0, rsp.) etc. Shape optimisation is connected with searching for optimal structural boundaries, whereas topology optimisation is mainly associated with the minimum mass (volume) problems taking into account the area occupied by a structure, its thickness and/or density of fibres in the construction.

Table 2. The form of the incidence matrix $a(i,j)$

1	0	0	0	0
0	0	1	1	0
0	1	0	0	0

Table 3. The physical sense of the terms in the incidence matrix

Optimisation Problem	Row	Column
Stacking sequence	Orientation	Ply Number
Shape	Position	Number of Key Point
Topology	Existence of FE	Number of FE

3.1   Shape optimisation

In the shape optimisation problem the required boundary curve is evaluated with the use of the Bezier splines – see Fig.1. Each keypoint (six in Fig.1) is represented by the radius (denoted by the position in Table 3). The radii are defined on a discrete set. Dependly on the Bezier splines construction the convexity of the curves may be prescribed in advance together with the required values of derivatives at the curve ends and other constraints (e.g. the total area under the curve).

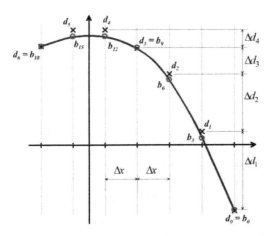

Figure1. Construction of a curve describing the required shape

3.2   Topology optimisation

Topology optimisation problems are based on the elimination of sets of FE from the initial mesh. The total length of the chromosome is composed of a finite number of parts, and each of them corresponds to a curve joining the base points. Two independent methods have been introduced: (1) the variation of the base points – Fig. 2 and (2) the elimination of domains enclosing the given set of curves – Fig.3. In the first case (1) the base points are connected with the use of straight lines and the lines constitute the row of FE having the prescribed properties. In the second case (2) the area between the curved lines is filled by generated FE.

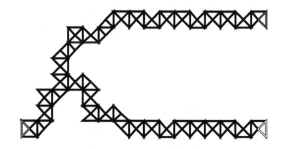

Figure 2. Variation of the base points

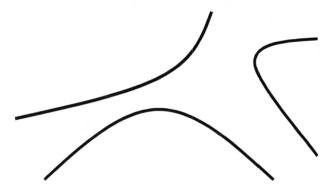

Figure 3. Elimination of domains

The curves enclosing the selected domains vary in the optimisation problems. They are described in the similar manner as for the shape optimisation problems via a finite number of the key points.

4. EFFECTIVENESS OF OPTIMISATION

4.1 Benchmarks

For shape and topology optimisation problems the effectiveness of the optimisation procedures and methods of coding have been determined using the known theoretical solutions given in a theoretical form.

4.1.1 Shape Optimisation

For shape optimisation problems the accuracy in the boundary curve evaluation is the most important problem. To verify it the following well-known problem have been solved: *to find the maximal area A surrounded by the curve with the fixed length $\tilde{L}$*. The problem has been formulated in the following way:

$$Max\left[Area \square pen * \left(\tilde{L} \square L\right)\right] \tag{3}$$

The results of the numerical solution are shown in Fig. 4. They demonstrate an excellent accuracy. The error between theory and numerical analysis is equal to 0.66% in the evaluation of the area A, and 0.11% - the length L.

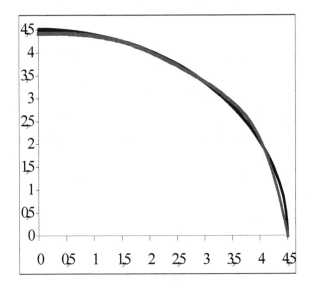

Figure 4. Comparison of numerical and theoretical results (red curve – analytical results)

4.1.2   Topology Optimisation

Let us consider a composite beam reinforced by short fibres, loaded at the center (Fig. 5 – three point bending test) and having a variable density of fibres $\rho_f(x)$ along the length.

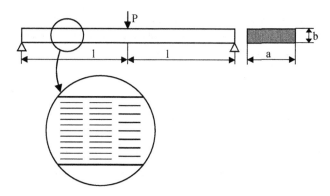

Figure 5. Bending of a simply supported beam having a variable fibre volume fraction distribution

The objective of the optimisation is following: *to find the minimal mass of the beam satisfying the equality constraint imposed on the displacement parameter and inequality (upper and lower bounds) constraint - on the fibre volume density fraction V_f* – the strict formulation is given by Banichuk et al. [3]. The optimisation problem deals with the topology optimisation since we are looking for the optimal material distribution. The results are presented in Fig. 6 and show a very good correlation between numerical and analytical studies.

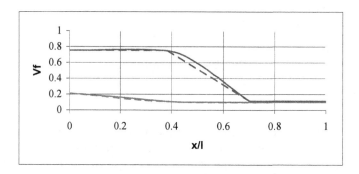

Figure 6. Comparison of analytical and numerical (a continuous line) results

4.2 Convergence of probabilistic algorithms

The similarity (identity) of optimal design for various probabilistic algorithms is understood in the sense of the convergence of the numerical results to the global optimum. The detailed analysis deals with the topology optimisation problem of a composite rectangular plate subjected to the action of a

shearing point load at the plate center. We are looking for the minimal mass
of the plate satifying the additional constraint conditions determined for the
values of allowable displacements and the effective stresses – see Ref [1].

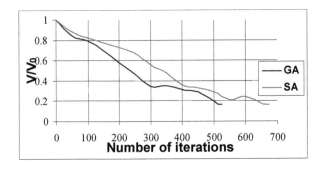

Figure 7. Variation of the base points

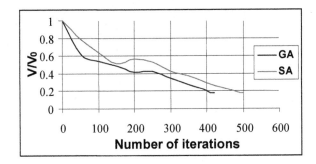

Figure 8. Elimination of domains

Figures 7 and 8 demonstrate the effectiveness of the optimisation
problems in the sense of number of iterations required in the analysis. The
obtained topologies are almost the same in the sense of the final volume
(area) of the structure both for the simulating annealing algorithms as well
as genetic algorithms. For all analysed numerically cases with the different
degrees of anisotropy in the composite plates the differences (of the value of
optimal volume) between different variants of GA and S.A. do not exceed
2%. However, the total number of iterations is always higher for problems
solved with the use of S.A. algorithm.

5. OPTIMISATION IN A FUZZY ENVIROMENT

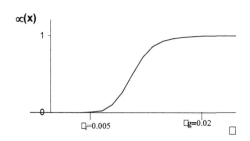

Figure 9. The possible form of the membership function $\propto$

A statistical distribution of mechanical properties is a standard phenomenon for composite materials. A fuzzy set analysis offers a powerful tool which may be applied in all problems where impreciseness, uncertainty, fuzziness play an important role and become significant in the evaluation of the output data understood in the sense of failure loads for composite structures.

For instance let us consider the experimental data characterizing the values of laminate allowable strains in tension which lie in the interval $[\varepsilon_l, \varepsilon_g]$. Using the fuzzy set approach it is possible to describe the gradual transition between the state of the lack of failure $\varepsilon < \varepsilon_l$ and the complete failure if $\varepsilon > \varepsilon_g$. It is described with the use of the membership function (see e.g. Fig. 9) where e.g. the statistical data may be used in the derivation of the particular membership function form.

To illustrate and demonstrate the sense of the introduced description of optimisation in a fuzzy enviroment the stacking sequence (topology) optimisation problem for rectangular bi-axially compressed and subjected to a uniform normal pressure multilayered composite plate have been solved. The objective function is given by eqn (1). We intend to maximize the loading parameter q. The objective is subjected to the set of six fuzzy constraints corresponding to the values of allowable strains in the longitudinal, perpendicular and transverse directions to fibres (the simplest FPF criterion in the form of the maximal strains) for tension and compression, independently. The form of the membership functions (six functions) is assumed to be identical to that plotted in Fig. 9. The different parameters β characterizing the slope of the curve have been analysed starting from 0 (the deterministic, crisp analysis) to 0.5 to describe various values of the fuzziness.

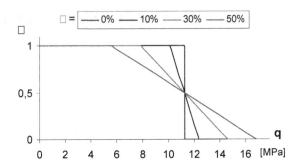

Figure 10. The fuzziness of the optimal loading parameter q

Then with the use of classical relations for deformations of plates one may express in the analytical form the relations between strains in each individual ply, external loads and fibre orientations. Applying genetic algorithms one may find the optimal fibre orientations in plies as the function of the load parameter and of the fuzziness related to the slope of the curve plotted in Fig. 9. The EVOLVER package have been also used in numerical computations. For instance, for the laminated structure made of 20 plies (symmetric, balanced laminates) the optimal fibre orientation is: $[90_2°,90_2°,90_2°,90_2°,90_2°,90_2°,90_2°,90_2°,90_2°,0_2°]_s$. The optimal fibre orientations are a function of material and geometrical parameters as well as of the total numbers of layers in the laminate. Figure 10 demonstrates the fuzziness of the loading parameter q for various parameters β and λ. In the analysed case the optimal stacking sequence written above is independent on the fuzzy constraints, however the maximal loading parameter may vary.

References

[1]   Muc, A., Gurba, W., Genetic algorithms and finite element analysis in optimisation of composite structures, *Composite Structures*, **54**, pp. 275-281, 2001.

[2]   Muc, A., Gurba, W., Optimization of volume for composite plated and shell structures, *Proc. 3rd Int.Conf. Thin-Walled Str.*, pp. 585-592, Elsevier, London., 2001.

[3]   Baniczuk, N. W., Iwanowa, S. Ju., Szaranjuk, A.W. , *Dynamics of structures. Analysis and Optimisation*, Nauka, Moscow, 1989 (in Russian).

GENETIC ALGORITHMS IN OPTIMISATION OF RESIN HARDENING TECHNOLOGICAL PROCESSES

A. Muc[1], P. Saj[2]

[1]*Institute of Mechanics & Machine Design, Cracow University of Technology, Kraków*
[2]*ABB Research Corporate Center, Kraków, Poland*

Abstract: For the RIM and RTM processes two optimisation problems have been formulated and solved. In the first case the quality of the manufactured product is evaluated by the spatial distribution of the degree of cure, whereas in the second case by the values of the residual strength arising during the chemical reactions and thermal stresses. The design variables represent temperatures and/or heating regime during technological processes.

Keywords: RIM process, RTM process, optimisation, genetic algorithms.

1. INTRODUCTION

Recently, the design of technological processes is conducted with the use of numerical modeling of them and then the experimental verification of the results. Numerical simulation of complex physical phenomena allows us to introduce and solve different optimisation problems. In the area of resin moulding various numerical methods (FDM, FEM, CVM) have been used to describe a progress and development of different physical processes. Such technological processes are especially difficult in numerical description due to complexity of physical modeling of them and on the other hand with respect to complex 2-D or 3-D forms of moulds and of products.

In the present paper the optimisation problems for two similar techno-logical processes will be discussed in details: the resin transfer moulding (RTM) through a porous media (composite materials) and the reactive injection moulding (RIM). The numerical simulation is conducted with the

287

T. Burczyński and A. Osyczka (eds),
IUTAM Symposium on Evolutionary Methods in Mechanics, 287–296.
© *2004 Kluwer Academic Publishers.*

use of the control volume element method – the package FLUENT. The physical models used in the analysis are discussed in details in Refs [1], [2].

2.   GOVERNING RELATIONS

For the non-isothermal process the fundamental set of equations should describe simultaneously different stages of the technological process, i.e.: (1) the behaviour of the fluid (the synthetic resin) during the filling of the mould coupled with (2) a heat transfer analysis from the mould walls and (3) the reactive nature and complex rheology of reactive materials (the viscosity and curing kinetics) during the gelation stage of the process. The Navier-Stokes equations are commonly used in modeling of a such class of problems, however, they are supplemented by a set of additional relations describing the viscosity and curing kinetics. In the case of the viscous Newtonian fluid the set of fundamental Navier-Stokes equations may be written in the following integral form:

$$\frac{\Box}{\Box t}\int_{\Box}\vec{W}d\Box \;+\; \oint_{\Box}\left(\vec{F}_c \,\Box\, \vec{F}_v\right)dS \;=\; \int_{\Box}\vec{Q}d\Box \tag{1}$$

or in the equivalent differential form:

$$\frac{\Box\vec{W}}{\Box t}+\Box\cdot\left(\vec{F}_c \,\Box\, \vec{F}_v\right)=\vec{Q} \tag{2}$$

where:

$$\vec{W}^{Tr}=\left[\Box,\Box V_j,\Box E\right],\quad \vec{F}_c^{Tr}=\left[\Box V,\Box V_i V_j + p\Box_{ij},\Box HV\right],\quad h=cT,$$

$$\vec{F}_v^{Tr}=\left[0,*_{ij},6_{i}\right],\quad \vec{Q}^{Tr}=\left[0,\Box f_i,\Box f_i V_i + q_h\right],\; 6_{i}=V_j *_{ij}+\Box\frac{\Box T}{\Box x_i},$$

$$H=h+\frac{1}{2}V_i V_i = E+\frac{p}{\Box}, \tag{3}$$

The above relations describe the exchange (flux) of mass, momentum and energy through the boundary $\partial\Omega$ of a control volume Ω, which is fixed in space. There are two flux vectors $\vec{F}_c, \vec{F}_v$. The first one is related to the convective transport of the quantities in the fluid. Usually it is termed vector of convective fluxes, although for the momentum and the energy equation it also includes the pressure terms $p\vec{n}$ where p denotes a static pressure and $\vec{n}$ is a unit, normal vector. $\vec{V}$ means the velocity vector having the components V_j. The second flux vector – vector of viscous fluxes, contains the viscous stresses $*_{ij}$ defined in the following way:

$$*_{ij} = \alpha\left(\frac{\partial V_i}{\partial x_j} + \frac{\partial V_j}{\partial x_i}\right) - \frac{2}{3}\alpha\frac{\partial V_l}{\partial x_l}\delta_{ij} \tag{4}$$

as well as the heat diffusion, where T is a temperature, λ is a thermal conductivity coefficient and α - molecular fluid viscosity coefficient. H is the total enthalpy, h means the enthalpy and f_i are components of the body forces whereas q_h denotes the time rate of the heat transfer per a unit mass. E denotes the total energy of a fluid per a unit mass.

The above general relations describing flow of the resin and the heat transfer have to be supplemented by the equations characterizing the rheokinetics of the resin in the sense of the viscosity relations:

$$\exists = \frac{\alpha}{\Box} = B\,exp\left(\frac{T_b}{T}\right)\left(\frac{\Box_g}{\Box_g - \Box\Box}\right)^{C_1 + C_2\Box\Box} \tag{5}$$

$$\frac{d\Box}{dt} = q_h = \left(k_1 + k_2\Box^m\right)\Box(1 - \Box\Box)^n, k_i = A_i e^{\frac{\Box E_i}{RT}}, i=1,2 \tag{6}$$

where : B, C_1, C_2, T_b – constants; T - the absolute temperature.; $\Box_g$= 0.65 (65% of a cured material); $\Box$- degree of cure at the given temperature T [%] ; m, n - constants; k_i - reaction rate constants; A_i- pre-expotential factors; E_i - activation energies; R - the universal gas constant. According to this model a degree of cure at time t is defined as: $\Box = G(t)/G_\Box$ where $G(t)$- the heat of reaction released at time t at a given temperature, and $G_\Box$ - the total heat of reaction at a sufficiently high temperature at which the resin is perfectly cured.

Dependly on the used model the coefficient k_1 may be identically equal to zero (the model Bogetti, Gillespie) [3] or not.

The set of governing equations (1)-(6) is supplemented by the boundary and initial conditions formulated at the resin inlet, at the boundaries of the mould and at the free surface of the resin filling the mould and additionally for the RTM processes by the Darcy flow rule:

$$v_i = \frac{[K]}{\exists}(\Box p)_i \tag{7}$$

describing the penetration of the resin through the porous medium of fibres.

3.   CODING OF THE DESIGN VARIABLES

Both technological processes discussed herein, i.e. the RIM and the RTM processes are conducted in the similar way presented in Fig.1.

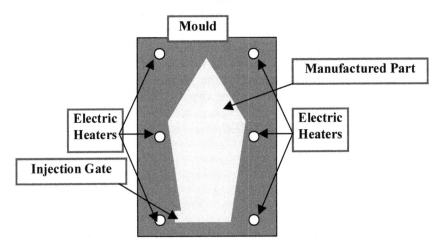

Figure 1. 2-D cross-section of the mould with manufactured part

The initially heated resin mixture is injected into the mould through the inlet gate – their number may be greater then one. In the part of the mould occupied by the manufactured part a porous media may exist that detemines unidirectional, 2-D or 3-D systems of fibres in the case of the RTM process and one or more obstacles in the macro scale (or even none of them) in the case of the RIM process. In the mathematical or numerical sense the difference between two processess results in the appearance of the additional set of equations (the Darcy flow rule – Eqn (7)) for the RTM process whereas for the RIM process the additional part inserted into the space occupied by the manufactured part in Fig.1 leads to the apperance of additional boundary conditions at the boundaries of obstacles. Then the whole mould is heated to the prescribed temperature and then consolidated and cooled. The temperature in the mould is controlled by a finite set of electric heaters running around the mould in 3-D space.

In optimisation problem the optimisation procedure and forms of design variables are mainly connected with the definition of the temperature regime – see Fig.2.

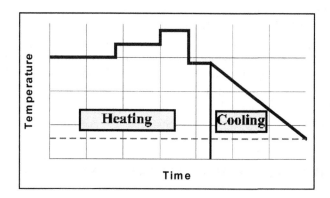

Figure 2. Heating and cooling regime of the manufactured parts

With the regard to optimisation problems dealing with the numerical analysis of the Navier-Stokes equations for technological processes one may distinguish two general groups of objectives:

— Shape optimisation of moulds resembling structural shape optimisation problems and in view of that FE packages are rather preferred since the numerical formulation of a such class of problems is well established,

— Optimisation of parameters or boundary/initial conditions of technological processes; in such an analysis the advantages in the application of the CVM are rather obvious.

Of course, in the second case the degree of cure α and the distribution of temperatures T inside the mould vary with boundary conditions (the number of inlet gates, values of pressures etc.), positions of electric heaters located outside the mould, the values of the temperatures at those points and on many mechanical and thermal parameters describing the cure kinetics of the problem. Each of them may be treated as design variables of the optimisation problem. In addition, it is possible to assume that design variables s are constant in time during the heating part of the thermal process, i.e.:

$$s_k(x_i,t) = s_k(x_i) \tag{8}$$

or they may vary in time but in a specific, prescribed way, i.e.:

$$s_k(x_i,t) = s_k(t\ ), \quad t \square [t_{p1},t_{p2}] \tag{9}$$

where k denotes the total number of design variables, whereas *p1* and *p2* means the number of time intervals in the assumed heating process – see Fig.2. The cooling phase of the thermal process (Fig.2) is analysed for the RTM process only.

The identical coding procedure is applied to the optimisation problems denoted above as 1 and 2 via the Bezier spline functions. However, design variables described by the Eqn (8) are used in the analysis of the RIM process, whereas the second relation (9) will be applied for the optimisation of the RTM processes.

4.   OBJECTIVES OF OPTIMISATION

4.1   The RIM Process

It is obvious that the heating of the mould in room or very low temperatures can reduce significantly or even eliminate the problems mentioned in the previous section. However, on the other hand, the quality of the product and its strength increases significantly for the curing process at high temperatures. Therefore, it is necessary to conduct the heating process at high temperatures but analysing both spatial as well as time distributions of the degree of cure α and these effects should be taken into account in the objective. Finally, the objectives of our optimisation problem can be formulated in the following way: to maximize

$$F = \sum_{i=1}^{I_h} \frac{\int_0^{t_k} \square_{Pi}}{t_k \square 100} \square \sum_{t_i=1}^{N_t} A \square pen \tag{10}$$

where : I_h – the total number of the control points, t_k – the total time of the curing process, $\square_{Pi}$ - the degree of curing at the control point P_i, t_i – the i-th time step, N_t – the total number of time steps in the numerical simulation, *pen* – the penalty coefficient, A – the value computed from the formulae:

$$A = \begin{cases} \dfrac{d\square}{ds_P} & \text{where} \quad \dfrac{d\square}{ds_P} > 0 \\[2mm] 0 & \text{where} \quad \dfrac{d\square}{ds_P} < 0 \end{cases} \tag{11}$$

4.2 The RTM Process

In the RTM process due to variety of chemical exothermal reactions and curing procedures residual stresses of non-mechanical origin can arise and finally may lead to the reduction of the load carrying capacity of the structure. Therefore the fundamental optimisation problem takes the following form: to minimize the residual stresses varying the time interval during the curing procedure – see Eqn (9), i.e.

$$Min\ \sigma(t) \tag{12}$$

using the appropriate physical model characterizing the non-mechanical strains and stresses – see Hahn et al. [4].

5. NUMERICAL MODELING

Each design variable given by Eqs (8) or (9) is a real number. It is discretised and represented as a binary string of a finite length. Increasing the length of the string it is possible to obtain the required accuracy. Then the genetic algorithm procedures conjugated with the FLUENT package are used to find the optimal solution. It should be emphasized that the fundamental problem lies in a good and correct numerical modeling of the physical problem. Therefore the numerical control volume model is build at the beginning of the numerical analysis in order to verify the accuracy of computations and then it is applied in the optimisation process.

6. OPTIMISATION RESULTS

To illustrate the effectiveness of the proposed method the analysis of the RIM process will be discussed herein. Figure 3 demonstrates the beginning of the process – the filling of the mould by a liquid, initially heated to the temperature 40^0 C epoxy resin as well as the position of the inlet gate. In the present analysis 3-D problem have been considered. It is assumed that the mould is heated along six points and their temperature are treated as design variables.

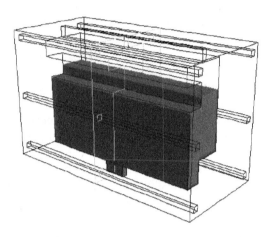

Figure 3. Filling of the mould by a liquid resin

Commonly, the heating of the resin is conducted at the constant temperature along the lines which is equal to 140^0 C. Figure 4 shows the final distribution of the degree of cure α at the end of the heating process.

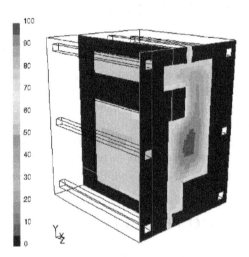

Figure 4. The distribution of the degree of cure α for the classical heating process at the temperature 140^0 C (the total time of the process 880 sec)

As it may be seen the gelation of the epoxy of the resin is almost constant inside the mould what is not very convenient since the new resin cannot be inserted into the mould and the air bubbles or voids may arise inside the manufactured part.

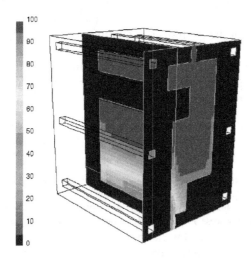

Figure 5. The optimal distribution of the degree of cure α (the total time of the process is now reduced to 815 sec)

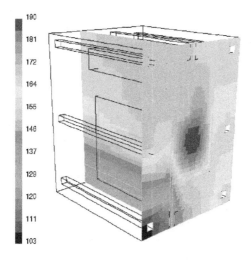

Figure 6. The optimal distribution of temperature around the mould at the end of the process

Conducting the optimisation analysis, where each temperature along the electric heaters lines is represented by the string having seven numbers one can obtain much better distribution of the resin (see Fig. 5). As it may be observed the gelation of the resin starts very far from the injection gate located at the lower left part of the picture (see Fig. 3) and finishes close to the gate. Such a situation is the best from the quality of the product point of view since the flow of the fresh resin is continuous during the whole curing process.

The optimal distributions of temperatures around the mould is shown in Fig.6. In addition, one can notice that the total time of the curing processes has been reduced but less then 10%.

References

[1]   Muc A., Saj P., Numerical optimisation of mould temperature regime in the RTM process, *Proc. ICCM/13*, Bejing, 2001 (CD 8 pages).

[2]   Muc A., Saj P., Numerical simulations and optimisation problems for reactive injection moulding process, *Proc. ECCM/2*, Kraków 2001 (CD 10 pages).

[3]   Bogetti, T.A. and Gillespie, J.W., „Processing induced stresses and deformation in thick – section thermosetting laminates", *CCM Reports*, Univ. Delawere, 1989

[4]   White , S.R. and Hahn, H. T., „Process modeling of composite materials: residual stress development during cure. Part 1. Model formulation", *Journal of Composite Materials*, Vol. 26, 1992

HYBRID EVOLUTIONARY ALGORITHMS IN OPTIMIZATION OF STRUCTURES UNDER DYNAMICAL LOADS

Piotr Orantek

Department for Strength of Materials and Computational Mechanics,
Silesian University of Technology, Konarskiego 18a, 44-100 Gliwice, Poland

Abstract: This paper is devoted to the application of hybrid evolutionary algorithm (HEA) in optimization of structures under dynamical loads. The HEA algorithms is a coupling of evolutionary and gradient algorithms, additionally the artificial neural network is used to control the selected parameters of this algorithm. The NURBS curves were used to model the shape of the structure. Three special types of chromosomes were tested to control the number of internal voids in the case topology optimization. The boundary element method [2] was need to compute the fitness function. The fitness function was expressed as the function depending on the displacements, stresses, mass, eigenfrequencies and compliance. Several tests were made [7], more interesting ones are presented in this paper.

Keywords: evolutionary algorithm, gradient algorithms, neural networks, boundary element methods, optimization

1.   INTRODUCTION

Both kinds of algorithms (evolutionary and gradient algorithms) were used separately so far, and when they were used skillfully, the results were satisfactory. However, in the case of more complex problems, the disadvantages of both attitudes were more evident. The alternative can be coupling the both algorithms into one, called the hybrid evolutionary algorithm. In this paper an idea of coupling the algorithms is presented. Moreover, the resultant algorithm is supported by the artificial neural network (ANN), defining the direction for work of the algorithm.

297

T. Burczyński and A. Osyczka (eds),
IUTAM Symposium on Evolutionary Methods in Mechanics, 297–308.
© 2004 Kluwer Academic Publishers.

2.   HYBRID EVOLUTIONARY ALGORITHM

A hybrid evolutionary algorithm is constructed as a coupling of on evolutionary algorithm [1][6] and the gradient method.

The special kind of mutation was introduced: a gradient mutation. This operator changes any chromosome $Ch(k)$:

$$[gen_1(k), gen_2(k), \ldots, gen_i(k), \ldots, gen_n(k)] \tag{1}$$

where x_i, $i = 1, 2, ..., n$, are genes.

The chromosome after mutation is given as:

$$[gen_1(k+1), gen_2(k+1), \ldots, gen_i(k+1), \ldots, gen_n(k+1)] \tag{2}$$

The new genes are:

$$gen_i(k+1) = \left\{ \begin{array}{ccc} gen_i(k) & gdy & l = 0 \\ gen_i(k) + \Delta gen_i(k) & gdy & l = 1 \end{array} \right.$$

where: $k = 0, 1, 2, ..., M$ - is the number of mutations, l - a random value which takes value $l = 0$ or $l = 1$. Correction values $\Delta gen_i(k)$ create vector $\Delta gen(k)$ which takes the form:

$$\Delta gen(k) = \alpha(k)\xi(gen(k)) \tag{3}$$

where: $\alpha(k)$ - a coefficient describing a step size, $\xi(gen(k))$ - a vector describing the direction of searching. Vector $\xi(gen(k))$ can be constructed on the ground of the knowledge of a gradient and a hessian. In the hybrid algorithm three kinds of the gradient mutation may be proposed. The formulas below describe how to determine the searching direction for different types of mutation:

■ for the steepest mutation:

$$\xi(gen(k)) = -\nabla f(gen(k)) \tag{4}$$

■ for the conjugate gradient mutation:

$$\xi(gen(k)) = -\nabla f(gen(k)) + \beta \xi(gen(k-1)) \tag{5}$$

where: β is the coefficient describing the influence of previous gradient, $\xi(k)$, $\xi(k-1)$ - conjugate directions.

■ for the variable metric gradient mutation:

$$\xi(gen(k)) = -D_k \nabla f(gen(k)) \qquad (6)$$

where: $D(k)$ is a hessian inverse matrix approximation computed on the base of the gradient.

The performed tests have shown, that presented attitude gives very positive results. The algorithm reached the global optimum faster thanks to the gradient operator. During testing such an algorithm it can be noticed that the modification of the parameters when the algorithm works can decrease the computation time. The first parameter is the population size. It can be controlled using the formula:

$$pop\_size_t = \alpha pop\_size_{t-1} \qquad (7)$$

where: t - generation, α - is a real value from interval $[0, 1]$. In this work received that a is a value from interval $[0.5, 1]$.

The second parameter is the iteration number of the gradient mutation. The gradient mutation was presented in the previous paper[??]. It has been decided that only the best individual is changed using the gradient mutation.

$$M_t = C \qquad (8)$$

It has been assumed, that considered parameters are modified using the history of the best individual's fitness function value.

In order to implement it the ANN has been applied. The ANN consists of 3 layers: input (10 neurons), hidden (4 neurons) and output (1 neuron) one. The input values are the appropriately transformed fitness functions of the selected individuals. The gamma value is the output value. The ANN was trained using 60 training vectors and verified for other 60. The selected parameters have been controlled using the following formulas:

$$pop\_size_t = \frac{2-\alpha}{2} pop\_size_{t-1} \qquad (9)$$

and

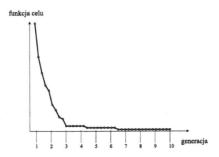

Figure 1.   The schema of global idea

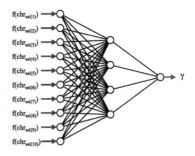

Figure 2.   The artificial neural network

$$M_t = int(C\alpha) \tag{10}$$

where: a - is a output value of the network, C- is a maximum number of mutation (in this work C=10).

3.   FORMULATION OF THE OPTIMIZATION PROBLEM IN DYNAMICS

A elastic body, which occupies a domain Ω bounded by a boundary Γ, is considered. Boundary conditions in the form of the displacement and traction fields and the initial conditions are prescribed [3][5].

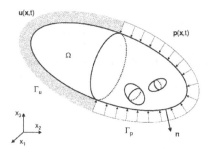

Figure 3.   The body under dynamical loads

One should find the optimal shape and topology of the boundary Γ by minimizing an objective functional:

$$\min_{x} Jo(\mathbf{x}) \tag{11}$$

with constraints:

$$J_\alpha(\mathbf{x}) = 0, \quad \alpha = 1, 2, ..., A \quad and \quad J_\beta(\mathbf{x}) \geq 0, \beta = 1, 2, ..., B \tag{12}$$

where: $\mathbf{x}$ is a vector of shape design parameters.

In work the fitness function was formulated as follows:

- eigenfrequencies functionals:

$$J = \omega_1, \tag{13}$$

$$J = \min_i |\omega_f - \omega_i|, \qquad i = 1, 2, 3, \tag{14}$$

$$J = \min_i |\omega_{i+1} - \omega_i|, \qquad i = 1, 2. \tag{15}$$

- compliance functional:

$$J = \frac{1}{T} \int_0^T \int_\Gamma \mathbf{p} \mathbf{u} d\Gamma dt \tag{16}$$

- stress functional:

$$J = \int_0^T \int_\Gamma \left[\frac{\sigma_{red}}{\sigma_0} \right]^n d\Gamma dt \tag{17}$$

where: σ_{red} - equivalent stress, σ_0 - admit stress, n - a parameter;

- displacement functional:

$$J = \int_0^T \int_\Gamma \left[\frac{\sqrt{\mathbf{u} \cdot \mathbf{u}}}{u_0} \right]^n d\Gamma dt \tag{18}$$

where: u_0 - admit displacement;

- mass functional:

$$J = \int_\Omega \rho d\Omega \tag{19}$$

4. MODELING OF THE SHAPE AND TOPOLOGY OF THE STRUCTURE

For the control of the shape of the body the NURBS curves were used. This approach can decrease the number of design variables.

The internal boundaries were modelled using the very well known shapes e.g. a circle, an ellipse, and the closed NURBS curves (for the 2-D problems) and a sphere, an ellipsoid and the closed NURBS surfaces.

Three special types of chromosomes were tested to control the number of internal voids in the case of topology optimization. The optimal combination of genes is presented in Fig 5.

The number of the internal voids depends of the condition imposed to the size of the void, e.g. radius for a circle.

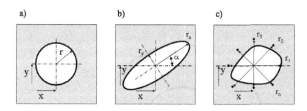

Figure 4. The kinds of the parametrization of the internal boundaries for the 2-D problems a) a circle, b) an elipse, c) a closed NURBS curve

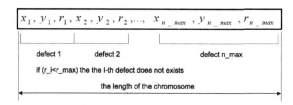

Figure 5. The optimal chromosome for the topology optimization

5. EXAMPLES

5.1 Numerical example 1.

A shape and topology optimization problem of the structure presented in Fig 6 is considered for criterion (15). The HEA searched the optimal topology and shape of the structure.

Figure 6. The structure before the optimization

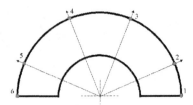

Figure 7.   The parametrization of the structure

The upper part of the boundary was modelled using the 6-point NURBS curve, additionally 4 circular holes could be introduced to the structure (Fig.7).

The population size of the HEA in the first generation (in the next generations the ANN controls the population size) was equal to 200. The number of generations was equal to 100. After the optimization the fitness function was decrased from 2226 to 2271 s^{-1}. The Fig. 8 shows the optimal topology and shape after the optimization.

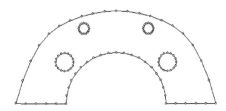

Figure 8.   The structure after the optimization

5.2   Numerical example 2.

A shape optimization problem of the structure presented in Fig. 9 is considered for criterion (17). The structure is loaded by force $F(t) = F_0 H(t); F_0 = 10kN$, and constrained on the left hole. The HEA searched the optimal shape of the structures. The outside part of the boundary was modelled using a polynomial and a circle (Fig.10).

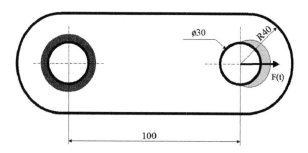

Figure 9. The structure before the optimization

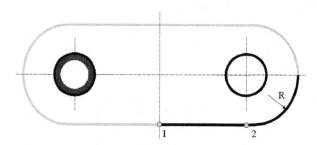

Figure 10. The parametrization of the structure

The population size of the HEA in the first generation (in the next generations the ANN controls the population size) was equal to 50. The number of generations was equal to 100. After the optimization the fitness function was decreased from 20382 to 5036 s^{-1}. The Fig. 11 shows the optimal shape after the optimization.

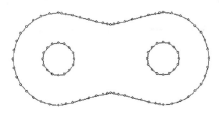

Figure 11.   The structure after the optimization

5.3   Numerical example 3.

A shape optimization problem of the structure presented in Fig. 12 is considered for criterion (19). The structure is loaded by force $F(t) = F_0 sin(\omega t); F_0 = 10kN$, and constrained on the lower part. The HEA searched the optimal shape of the structures.

The free part of the boundary was modelled using the 8-point NURBS surface. The population size of the HEA in first generation (in the next generations the ANN controls the population size) was equal to 50. The number of generation was equal to 400. After the optimization the fitness function was decreased from 0.98 to 0.87 kg.

6.   CONCLUSIONS

The time of evolutionary optimization can be effectively shortened by the application of the application of the artificial neural network (by 50%). Performed tests show that this approach is resistant for the kind of fitness functions.

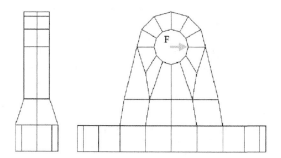

Figure 12.   The structure before the optimization

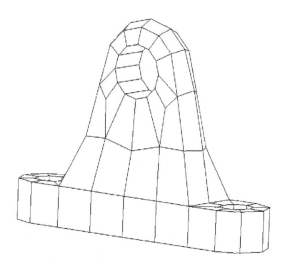

Figure 13.   The structure after the optimization

ACKNOWLEDGEMENT

The support from KBN grant no. 4T11F00822 is gratefully acknowledged.

References

[1] J. Arabas. *Wykłady z algorytmów ewolucyjnych.* WNT, 2001. (in Polish)

[2] T. Burczyński. *Metoda elementów brzegowych w mechanice.* WNT, Warszawa 1995. (in Polish)

[3] C.A. Brebbia, J. Dominguez. *Boundary Elements-An Introductory Course.* McGraw-Hill, 1992.

[4] E. Kita, H. Tani, Shape optimization of continuum structures by genetic algorithm and boundary element method, *Engineering Analysis with Boundary Elements*, Vol. **19** (1977).

[5] M. Kleiber (ed.) *Handbook of Computational Mechanics,* Springer (1998).

[6] Z. Michalewicz, *Genetic Algorithms + Data Structures = Evolutionary Programs.* Springer-Verlag, AI Series, New York (1992).

[7] P. Orantek. *Application of the hybrid algorithms in optimization and identification problems for dynamic structures.* PhD Thesis, Silesian University of Technology, Gliwice 2002. (in Polish)

EVOLUTIONARY OPTIMIZATION SYSTEM (EOS) FOR DESIGN AUTOMATION

Andrzej Osyczka[1], Stanisław Krenich[2], Joanna Krzystek[2], Jacek Habel[2]
[1]*AGH University of Science and Technology, Department of Management,Gramatyka Str. 10, 30-067 Cracow, Poland, email: osyczka@mech.pk.edu.pl*
[2]*Department of Mechanical Engineering, Cracow University of Technology, 31-864 Krakow Al. Jana Pawla II 37, POLAND. Email: osyczka@mech.pk.edu.pl, krenich@mech.pk.edu.pl, habel@mech.pk.edu.pl,krzystek@mech.pk.edu.pl*

Abstract: In the paper the software package called Evolutionary Optimization System (EOS) is presented. This package contains several single and multicriteria optimization methods based on evolutionary algorithms. EOS is coded in the ANSI C language and it is a user-friendly computer program with the window structure. The system can be used to solve some design automation problems. In the paper the spring design automation problems are discussed and these problems show clearly that using EOS we can obtain automatically the optimum design of the springs giving only the basic data.

Keywords: evolutionary algorithms, evolutionary optimization system, spring design automation.

1. INTRODUCTION

Many design optimization problems are modelled by means of nonlinear programming (see for example [3,5]). The use of conventional optimization methods, such as gradient based methods, direct search methods etc., to solve these problems is limited. Recently genetic and evolutionary algorithms are widely used to solve complicated design optimization

309

T. Burczyński and A. Osyczka (eds),
IUTAM Symposium on Evolutionary Methods in Mechanics, 309–320.
© 2004 Kluwer Academic Publishers.

problems (see [2,4]). In the paper the software package called Evolutionary Optimization System (EOS) is presented. EOS is designed to solve single and multicriteria optimization problems for nonlinear programming problems, i.e. for the problems formulated as follows:

Find vector of decision variables

$$\mathbf{x}^* = [x_1{}^*, x_2{}^*, ..., x_I{}^*] \tag{1}$$

which will satisfy the K inequality constraints:

$$g_k(\mathbf{x}) \square 0 \quad \text{for } k = 1, 2, ..., K \tag{2}$$

and the M equality constraints:

$$h_m(\mathbf{x}) = 0 \quad \text{for } m = 1, 2, ..., M \tag{3}$$

and optimize the vector of objective functions:

$$\mathbf{f}(\mathbf{x}^*) = [f_1(\mathbf{x}), f_2(\mathbf{x}), ..., f_N(\mathbf{x})] \tag{4}$$

For single criterion optimization problems instead of the vector function $\mathbf{f}(\mathbf{x})$ we have the scalar function $f(\mathbf{x})$ which is to be minimized.

In this paper nonlinear programming modelling is used to solve design automation problems for some machine elements.

2. ARCHITECTURE OF EOS

Figure 1 shows the architecture of EOS. The system consists of three elements: windowed shell, evolutionary algorithm (EA) kernel and GNU – C language compiler.

EA system kernel is supplied as the object file (optimize.o), which is compiled using the public domain GNU – C compiler. The user prepares his optimization problem as C file (userfunc.c). This file should be compiled into the object form (userfunc.o) and than both object files are linked into the executable form (ea.exe). These tasks are performed automatically by EOS windowed shell (options from 7 to 9 – see Figure 1).

The windowed shell of EOS enables the user project management, i.e., creating and recording new projects and editing the existing ones. All projects are stored in separated files.

EOS windowed shell has editing fields, which are responsible for:

– Making the description of the optimization problem.

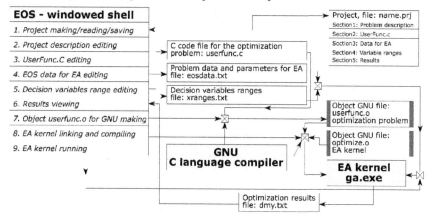

Figure 1. The interactive optimization system EOS architecture

– Creating the user function C source code.
– Introducing the input data for the optimization problem, evolutionary algorithm methods and additional data.
– Defining decision variable ranges.
– Storing optimization results.

3. GENERAL DESCRIPTION OF EA KERNEL SYSTEM

The general diagram of the EA kernel system is presented on Figure 2. The system is equipped with the following evolutionary methods:

1. For single criterion optimization:
a) Proportional selection method with the penalty function.
b) Bicriterion method
c) Simple tournament selection method with the penalty function.
d) Constraint tournament selection method.
2. For multicriteria optimization
a) Simple distance method.
b) Pareto set distribution method.
c) Constraint tournament selection method.

For multicriteria optimization problems the indiscernibility interval method can be used for selecting a representative subset of Pareto optimal solutions after running the assumed number of generation.

For both single and multicriteria optimization methods the following models can be solved:

a) with continuous decision variables,

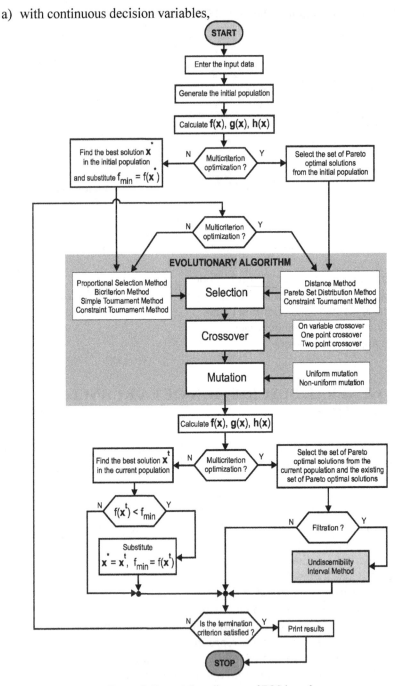

Figure 2. General flow diagram of EOS kernel.

b) with integer decision variables,
c) with discrete decision variables,
d) with mixed continuous – integer decision variables,
e) with mixed continuous – discrete decision variables.
   In EOS chromosomes can have:
a) binary representation,
b) real number representation,
c) Gray coding representation.
   Crossover operations can be performed as follows:
a) one point crossover,
b) two point crossover,
c) variable point crossover.
   Mutation operations can be performed as follows:
a) uniform mutation,
b) non-uniform mutation for binary representation and for real number representation.

4.   WINDOW STRUCTURE OF EOS

EOS is coded in the ANSI C language and it is a user-friendly computer program with the window structure. Some examples of the windows will be presented below.

In Figure 3 the window for the user supplied function, called *nonpro* function, is shown. This function is responsible for the introduction of an optimization problem into EOS. Its structure creates also the possibilities for design automation processes. The instructions under different cases of this function have the following meanings:

a) Under case 1 the basic problem dependent data are read and the additional data are calculated. Data base or expert systems can be implemented here.
b) Under case 2,3 and 4 an optimization problem is introduced into EOS by means of the description of objective functions, inequality and equality constraints.
c) Under case 5 the problem dependent results can be calculated and presented to the designer. In design automation processes this case can be used to create the file for obtaining the final draw of the design element or even for creating the file which is used for manufacturing process of this element.

Another window of EOS is presented in Figure 4. This window is responsible for introduction of the data while running EOS.

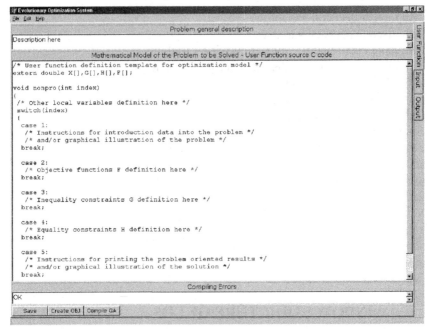

Figure 3. EOS screenshot – user function definition page

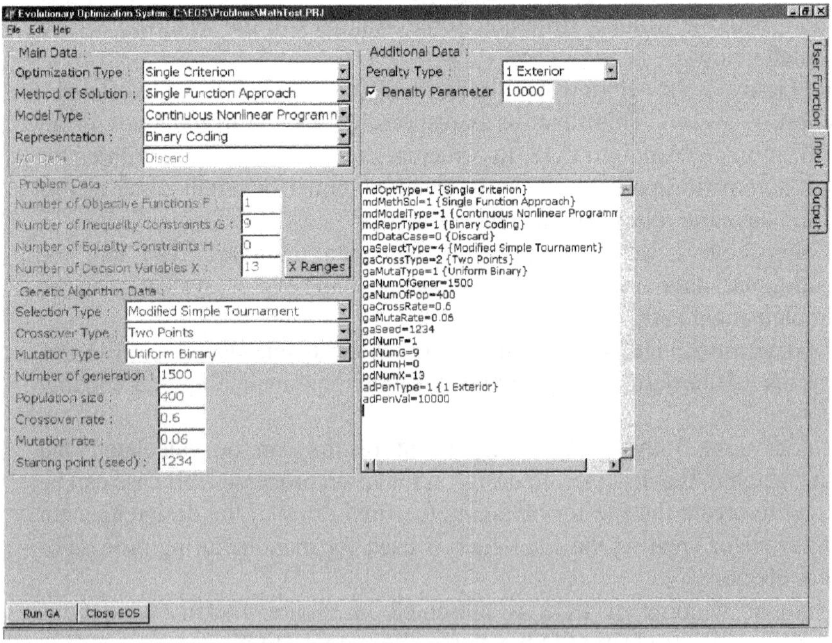

Figure 4. EOS screenshot – Input data page

Figure 5. EOS screenshot – output results page

Finally the window with the results of the optimization process is presented in Figure 5. All results from this window are stored in the file, which can be printed after the session with EOS.

5. METHODS OF DESIGN AUTOMATION

To solve design automation problem two methods can be used:

Method 1

This approach is based on two models with the respective computer programs.

Model 1: This model should find for the given input parameters an acceptable solution, i.e., a solution that satisfies constraints imposed on an element or an assembly.

Model 2: This model describes the optimization problem, i.e. the relations between the given input parameters, decision variables, constraints and the objective function. For Model 1 the method finds automatically an acceptable solution which is used to establish lower and upper bounds on the decision variables. Then the optimization process is carried out using those bounds employing Model 2.

Method 2

For some problems using the constraint tournament selection method the optimal design can be found using very wide ranges of the decision variables. These ranges can cover all possible dimensions for all designs we want to optimize. Thus, the proposed method is very simply and uses only Model 2 for which the evolutionary algorithm based method generates the solution within the following bounds:

$$x_i^l \ \Box \ x_i \ \Box \ x_i^u \text{ for } i = 1,2.....,I; \tag{5}$$

where: x_i^l - the smallest possible value of the i–th decision variable.

x_i^u - the greatest possible value of the i–th decision variable.

Running the constraint tournament selection method using the above bounds an optimal solution can be found automatically for any design. The method requires only the basic data to be introduced.

6.   SPRING DESIGN AUTOMATION

6.1   Optimization model of the helical spring

Let us consider design automation of coil compression springs the scheme of which is presented in Figure 6.

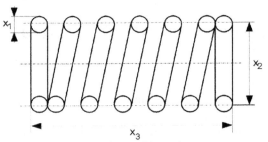

Figure 6. Scheme of the spring

Optimization model can be presented as follows. The vector of decision variables is:

$$\mathbf{x} = \left[x_1, x_2, x_3, x_4\right]^T, \text{ where:}$$

x_1 – wire diameter of the spring [mm],

x_2 – meancoil diameter of the spring [mm],
x_3 – length of the spring [mm],
x_4 – number of active coils [–].

The objective function is the volume of the spring, which can be expressed as follows:

$$f(\mathbf{x}) = \frac{\pi}{4} x_1^2 \sqrt{2 \Box x_2^2 x_4^2 + x_3^2} + \frac{\pi^2}{4} x_1^2 x_2 \tag{6}$$

Formulas for the constraints are based on Polish Standard PN–85/M–80701–3. Due to page limitation the constraints are described only verbally:
a) shear stress constraint,
b) stiffness of the spring constraint,
c) clearance between coils constraint,
d) buckling constraint,
e) geometric constraints.

The optimization model is considered as a discrete one, with the following sets of possible values of decision variables:

$X_1 = \{0.5, 0.63, 0.8,..., 6.3, 8.0, 10.0\}$,
$X_2 = \{1,2,3,4,...,60,61,62,...,300\}$,
$X_3 = \{1,2,3,...,50,51,52,...,600\}$,
$X_4 = \{1.5,2.5,...,49.5\}$

These sets cover almost all possible spring designs. The EOS can find the optimal solution automatically using only the following basic data:
a) material of the spring,
b) compression force,
c) stiffness or deflection of the spring,
d) type of the spring (running/no running).

After solving the optimization problem using the constraint tournament selection method the optimal design of the spring can be obtained automatically.

Numerical Example

Let us consider example of spring design. The input data taken from [1]:
a) Material of the spring - 45S,
b) Compression force - 1850 [N],
c) Stiffness of the spring - 20.55 [N/mm],
d) Type of the spring – non running

The results obtained using the automation design method are:

$$f(\mathbf{x}) = 84\,989.59, \quad \mathbf{x} = [8.0, 63.0, 159.0, 7.5]^T$$

Using these results EOS creates automatically the file for AutoCAD and then the final drawing of the spring is created. This drawing is shown in Figure 7.

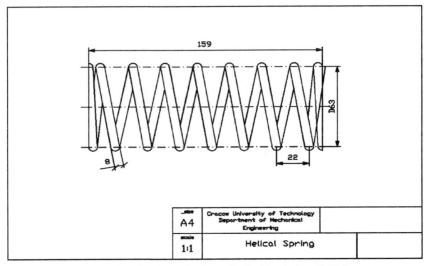

Figure 7. Engineering draws for the optimal spring in AutoCAD format.

6.2   Concentric spring design automation

Concentric spring design automation problem is considered as a second problem. The scheme of the spring the scheme of is presented in Figure 8. The optimization model was build on the basis of formulas from [4] and [6].

The optimization model is as follows:

Decision variables:

$$\mathbf{x} = [x_1, x_2, x_3, x_4, x_5, x_6, x_7]^T, \text{ where:}$$

x_1 – wire diameter of the outer spring [mm],
x_2 – meancoil diameter of the outer spring [mm],
x_3 – number of active coils of the outer spring [–],
x_4 – wire diameter of the inner spring [mm],
x_5 – meancoil diameter of the inner spring [mm],
x_6 – number of active coils of the inner spring [–],
x_7 – length of the spring system [mm].

The problem is finding dimensions of concentric springs, which satisfy constraints and minimize function of volume of the springs.

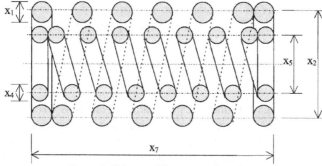

Figure 8. Scheme of the concentric springs

$$f(\mathbf{x}) = \frac{1}{4}\left(x_1^2 \sqrt{\pi^2 x_2^2 x_3^2 + (x_7 \pi x_1)^2} + x_4^2 \sqrt{\pi^2 x_5^2 x_6^2 + (x_7 \pi x_4)^2} \right) +$$
$$+ \frac{\pi^2}{4}\left(x_1^2 \pi x_2 + x_4^2 \pi x_5 \right)$$

(7)

Constraints:

Due to page limitation the constraints are described verbally: shear stress constraint, stiffness of the spring constraint, clearance between coils constraint, buckling constraint, geometric constraints, total stiffness of the spring, dependencies between diameters of both springs, difference between the safety factors of strength fatigue of both springs.

Numerical Example

The optimization problem is considered as a discrete programming problem with the following sets of discrete values of the decision variables:

X_1 = {0.5, 0.63, 0.8,..., 6.3, 8.0, 10.0 },
X_2 = {1,2,3,4,...,60,61,62,...,300},
X_3 = {1.5,2.5,...,49.5},
X_4 = {0.5, 0.63, 0.8,..., 6.3, 8.0, 10.0 },
X_5 = {1,2,3,4,...,60,61,62,...,300},
X_6 = {1.5,2.5,...,49.5},
X_7 = {1,2,3,...,50,51,52,...,600},

The input data for optimization process are as follows:
a) Load P = 1500 [N],
b) Spring constant c = 150 [N/mm],
c) Material of the spring – toughened spring steel 5HG.

The results of automation of design of the concentric springs are as follows:

$$f(\mathbf{x}) = 41\ 667.74, \quad \mathbf{x} = [10.0, 49.0, 1.5, 7.5, 30.0, 2.5, 46.0]^T$$

From this example it is clear that the proposed method can find automatically the solution for more complicated spring design problems.

7.   CONCLUSIONS

In the paper the Evolutionary Optimization System is presented and advantages of using the system in design automation processes are shown. These advantages refer to the process of creating the data for the optimization procedure, which generates the optimal solution automatically, and to the process of obtaining the data for the graphical illustration of the solution as well as creating the data for the manufacturing process.

In the paper single criterion optimization models are considered. In some design automation problems a multicriteria optimization approach should be used. Such approach creates difficulties, which refer to the decision making problem. This will be the subject of further investigations.

ACKNOWLEDGMENTS

This study was sponsored by Polish Research Committee, under the Grant No. 7 T07A 007 19.

References

[1]   Branowski B.: *Sprężyny metalowe*. In Polish, (Metal Springs), Wydawnictwo Naukowe PWN, Warsaw, 1997.

[2]   Gen, M. P. and Cheng, R.: *Genetic Algorithms and Engineering Design*. John Wiley and Sons, Inc., New York, 1997.

[3]   Haftka, R. T. and Kamat, P. M.: *Elements of Structure Optimization*. Martinus Nijhoff Publishers, The Hague, Boston, Lancaster, 1985.

[4]   Osyczka A.: Evolutionary Algorithms for Single and Multicriteria Design Optimization, Springer Verlag Phisica, Heidelberg, Berlin, 2002.

[5]   Papalambros, P. Y. and Wilde, D. J.: *Principles of Optimal Modeling and Computation*, Cambridge University Press, 1988.

[6]   PN–85/M–80701–3, Polish Standard. Sprężyny śrubowe walcowe z drutów lub prętów okrągłych. Sprężyny naciskowe, obliczenie i konstrukcja.

EVOLUTIONARY METHOD FOR A UNIVERSAL MOTOR GEOMETRY OPTIMIZATION

A new automated design approach

GREGOR PAPA, BARBARA KOROUŠIĆ-SELJAK
Computer Systems Dept., Jožef Stefan Institute, Jamova c. 39, SI-1000 Ljubljana, Slovenia

Abstract: A new designing approach for the rotor and the stator of a universal motor for home appliances is presented in this paper. It is based on a simple and efficient genetic algorithm. The aim is to optimize the geometric parameters of the rotor/stator lamination in an automated way to reduce the main motor's power losses. With this design procedure the motor's efficiency has been significantly improved.

Keywords: evolutionary optimization, universal motor, design automation

1. INTRODUCTION

Most home appliances with small motors, such as vacuum cleaners or mixers, are driven by a universal motor. Because of the widespread use of a universal motor, it is very important that the energy consumption of the motor (input power) is as low as possible, while the appliance still satisfies the needs of a user (output power). Hence, the technical quality of a universal motor can be expressed as efficiency, which depends on various losses.

The main losses, i.e., the iron and the copper losses can be reduced by optimization of rotor and stator geometry. Because of a high magnetic saturation of iron in universal motors, this is a highly non-linear problem. Due to the complexity of the solution originating in the non-linearity of the problem, a genetic algorithm (GA) [1] was used for optimization of rotor and stator geometry.

321

T. Burczyński and A. Osyczka (eds),
IUTAM Symposium on Evolutionary Methods in Mechanics, 321–330.
© *2004 Kluwer Academic Publishers.*

2. ROTOR AND STATOR GEOMETRY

The geometry of the rotor and the stator was defined parametrically. Some of the parameters were invariable, and were not altered due to the technological aspects. Only the dimensions of the variable parameters that were mutually independent (like the external radius of the rotor, the rotor pole width, the rotor jag diameter, the stator jag radius, etc.) were varied. In addition to the above-mentioned independent variables there are also dependent variables, which were defined from independent ones. All these dimension parameters are shown in Fig.1.

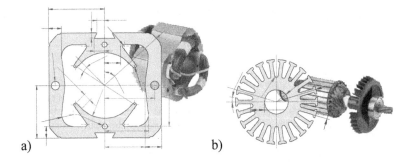

a)   b)

Figure 1. Geometric parameters: a) stator, b) rotor

3. DESIGN PROCEDURE

According to the conventional, i.e. direct, motor design procedure the initial estimation of the rotor and the stator geometry is based on experience. The appropriateness of a newly defined motor geometry is usually analyzed by means of a numerical simulation of the electromagnetic field. In our case, the analysis is performed by commercial software ANSYS [2], which applies a finite-element method with automatic finite-element mesh generation. In case the results of numerical simulation show an inconvenient electromagnetic field structure, the direct design procedure is repeated until the motor geometry is optimized.

The new motor design procedure described in this paper is a direct system also, but it is based on the GA. This stochastic process provides a robust and yet flexible search in the wide and complex space of the problem solutions in order to find an optimal global solution in a short time. The concept of the applied design system can roughly be explained as follows: the GA provides a set of problem solutions (i.e., different configurations of rotor and stator independent geometrical parameters). In order to enable the calculation of a fitness value, each geometrical configuration is analyzed using the finite-element program. After the calculation of the fitness, reproduction of individuals and application of genetic operators on a new population are done. The GA repeats this procedure until a predefined number of iterations have been accomplished.

4.   CALCULATION OF LOSSES

The finite-element program is used to calculate a fitness values of each solution, where fitness corresponds to iron and copper losses in the stator and the rotor of the motor.

4.1   Calculation of copper losses

The copper losses are the joule losses in the stator and in the rotor windings (stator and rotor slots):

$$J = \frac{I \square N}{A} \tag{1}$$

$$P_{CuSlot} = I^2 \square R = J^2 \square A \square \square l_{turn} \tag{2}$$

where J is current density, I is current, N is number of turns, A is slot area, $\square$ is copper specific resistance and l_{turn} is length of winding turn.

The overall copper losses are as follows:

$$P_{Cu} = \sum_i P_{CuSlot,i} \tag{3}$$

where i stands for each slot.

4.2   Calculation of iron losses

The iron losses include the hysteresis losses and the eddy-current losses, primarily in the armature core and in the saturated parts of the stator core. Because of the non-linear magnetic characteristic, the iron losses calculation is less exact. An empirical formula for the hysteresis loss is defined as follows:

$$P_h = k_h \Box B^n \Box f \Box m \tag{4}$$

where k_h is hysteresis material constant at 50 Hz, B is maximum magnetic flux density, f is frequency, m is mass, n is exponent between 1.6 and 2.0 (material dependent).

The eddy current loss is defined as follows:

$$P_e = k_e \Box B^2 \Box f^2 \Box m \tag{5}$$

where k_e is eddy-current material constant at 50 Hz.

The frequency of magnetic field density in a stator is 50 Hz (frequency of power supply), while in a rotor the frequency is much higher and depends on the motor speed. Since the eddy current losses depend on the square of the frequency, and the hysteresis losses increase linearly with the frequency, the main rotor iron losses are the eddy current losses. On the other hand, these losses depend on the square of the flux density. If the hysteresis losses in the rotor are neglected, the rotor iron losses can be calculated using (5) where B^2 is obtained from the finite-element solution for each node. Consequently, the overall B^2 of a rotor is calculated as an average of B^2 of all rotor nodes.

When calculating stator losses, hysteresis loss and eddy current loss must be summed up. Additionally, the factor n must be roughly estimated. Since the overall stator losses are approximately five times smaller than the rotor losses, even a non-exact estimation of a factor n does not have a significant influence on the overall motor iron losses calculation. Consequently, the iron losses of a motor can be expressed by the following equation:

$$P_{Fe} = k_e \Box B^2 \Box f_{rot}^2 \Box m_{rot} + k_e \Box B^2 \Box f_{stat}^2 \Box m_{stat} + k_h \Box B^2 \Box f_{stat} \Box m_{stat} \tag{6}$$

The sum of the iron and the copper losses P_{CuFe} represents a so-called objective function for the GA.

$$P_{CuFe} = P_{Fe} + P_{Cu} \tag{7}$$

4.3   Calculation of other losses and efficiency

Beside the iron and the copper losses, three additional types of losses also take place in a universal motor, where: P_{Brush} are brush losses, P_{Vent} are ventilation losses and P_{Frict} are friction losses. All three types of losses mainly depend on the motor speed. Since the motor speed is equal for all solutions, these losses are considered constant in our analysis. Considering all losses, the overall efficiency of a universal motor is defined as follows:

$$\eta = \frac{P_2}{P_2 + P_{Brush} + P_{Vent} + P_{Frict} + P_{CuFe}} \tag{8}$$

where P_2 is output power. Consequently, the goal of the present investigation is to maximize this efficiency.

The finite-element program needs the following data in order to solve the electromagnetic problem:
- well defined geometry of the analyzed element;
- appropriate finite-element mesh;
- material properties (iron B-H function, copper specific resistance);
- density of the electric current in the conductor area.

The result of the solution is a magnetic vector potential on every node of the finite-element mesh. From this potential, the values of flux density, field strength, magnetic energy and electromagnetic torque can be calculated. The output power of a motor is a product of the electromagnetic torque and the angular velocity.

5. GENETIC ALGORITHM

Traditional search and optimization methods are slow in finding the solution in a complex problem's search space. For this reason, we decided to apply the GA to increase the efficiency of a universal motor with respect to the geometry of its rotor and stator unit. The other reason was that the problem is a non-linear one. The GA is based on a heuristic method, which requires little information to search effectively in a large search space.

The GA codes parameters of the problem's search space as finite-length strings over some finite alphabet. It works with a coding of the parameter set, not the parameters themselves. The algorithm employs an initial population of strings, which evolve to the next generation by probabilistic transition rules such as selection, crossover and mutation. The objective function evaluates the quality (fitness) of solutions coded as strings. This information is used to perform an effective search for better solutions. There is no need of other auxiliary knowledge. The GA tends to take advantage of the fittest solutions by giving them greater weight, and concentrating the search in the regions of the search space with likely improvement.

The GA is different from the traditional techniques because of its intrinsic parallelism (in evaluation function, selections) that allows working from a broad database of solutions in the search space simultaneously, climbing many peaks in parallel. Thus, the risk of converging to a local optimum is low. The random decisions made in the GA can be modeled using Markov chain analysis to show that each finite GA will always converge to its global optimum region [3]. In spite of its simplicity, the GA has proved to be an efficient method for solving various optimization and classification problems, in areas ranging from economics and game theory to control system design [4,5].

Putting the new rotor and stator design procedure into practice the primary work was to define the encoding method of solution candidates, the genetic operators and the termination criteria. Also a method of evaluating the relative performance of solution candidates for identifying the better solutions had to be selected.

5.1   Encoding

Parameters of the problem's search space were coded as strings, i.e. chromosomes, over the alphabet 7 of real values. Using a symbolic presentation of a string with 11 characteristics (rotor and stator independent geometric parameters) gives:

$$S = s_1 s_2 s_3 s_4 s_5 s_6 s_7 s_8 s_9 s_{10} s_{11}$$

Here, each of the s_i represents a single stator or rotor geometrical parameter, where each characteristic may take on a real value from 7 .

The rotor and the stator independent geometric parameters of an existing universal motor were used to form a starting string, which was reproduced $(n-1)$-times to generate an initial population of n-strings. A random value distributed linearly on $\pm\square$ was added to each characteristic value of the starting string to define a reproduced string. The entire set of strings upon which the GA operated was called a population.

5.2   Genetic operators

To evolve the best solution candidate, the GA employed the genetic operators of selection, crossover and mutation for manipulating the strings in a population. The GA used these operators to combine the strings of the population in different arrangements, seeking a string that maximizes the objective function. This combination of strings resulted in a new population.

The first genetic operator used by the GA for creating a new generation was selection. To create two offspring two strings had to be selected from the current population as parents. Most fit strings were selected for reproduction. We had applied the elitism strategy, where a randomly selected number of least-fit members of the current population were interchanged with the equal number of the best-ranked strings.

Crossover proceeded in two steps. First, strings were mated randomly, using a given probability p_c to pair off the couples. Second, mated string couples crossed over, using a random probability to select the one-point crossing sites. An integer position k was selected between 1 and the string length less one $[1,l-1]$. Swapping all characteristic values between the positions $k+1$ and l inclusively created two new strings. For example, considering strings A and B and choosing a random number $k=4$. The resulting crossover yielded two new strings A' and B':

$$A = a_1a_2a_3a_4 \mid a_5a_6a_7a_8a_9a_{10}a_{11} \qquad A' = a_1a_2a_3a_4 \mid b_5b_6b_7b_8b_9b_{10}b_{11}$$
$$\Rightarrow$$
$$B = b_1b_2b_3b_4 \mid b_5b_6b_7b_8b_9b_{10}b_{11} \qquad B' = b_1b_2b_3b_4 \mid a_5a_6a_7a_8a_9a_{10}a_{11}$$

Moreover, we might use a constant probability p_r to select a case, in which the values of the swapped characteristics were calculated as a mean (average) value of the parent characteristic values.

$$a_{i_{new}} = b_{i_{new}} = \frac{a_i + b_i}{2} \tag{9}$$

While the first crossover approach assured that the child solutions preserved the 'genetic material' from both parents, the second one helped to seek for other solutions near to solutions appeared to be good.

Mutation was a process by which strings resulting from selection and crossover were perturbed. It served to create random diversity in the population. Each string was subjected to the mutation operator. Mutation was performed on characteristic-by-characteristic basis, each characteristic mutating with a probability p_m. However, since a high mutation rate resulted in a random walk through the GA search space, p_m had to be chosen to be somewhat low. There was also a possibility of annealing the mutation rate, where p_m was decreasing linearly with each new population. Namely, we assumed that each new population generally was more fit than the previous one. Such an approach was used to overcome a possible disruptive effect of mutation, and to speed up the convergence of the GA to the optimal solution in the final stages of the optimization.

5.3   Fitness evaluation

Following selection, reproduction, crossover, and mutation, the new population was ready to be evaluated. Therefore, each new string created by the GA was decoded into a set of rotor and stator geometrical parameters, and its fitness was estimated running the ANSYS. First, a finite-element numerical simulation was performed. Then, the iron and the copper power losses were calculated. Their sum (7) corresponded to the solution's fitness.

Additional criteria for the solution evaluation was the price of the material needed to make a motor. This criterion was considered through the outer dimensions of the lamination. Since we wanted to keep the amount of material constant regarding the initial design we also made some optimizations with fixed outer dimensions.

5.4 Termination criteria

The GA operated repetitively, with an idea that, on average, solutions of the population defining the current generation had to be as good (or better) at maximizing the fitness function as those of the previous generation. When a certain number of populations had been generated and evaluated, the system was assumed to be in a non-converging state. This criterion was a 'time-out' approach. The fittest member of the current generation at the time the GA terminated was taken to be the solution of the design problem.

5.5 Parameter settings

Finding good settings for parameters of the GA that work for the problem was not a trivial task. Robust parameter settings had to be found for population size, number of generations, selection criteria and genetic operator probabilities:
- If the population size was too small, the GA converged too quickly to a local optimal solution and might not find the best solution. On the other hand, large population required long time to converge to a region of the search space with significant improvement.
- Applying the elitism strategy fitter solutions had greater chance of reproducing. But when the number of least-fit solutions to be exchanged with best-fit ones (the selection criteria) was too high, the GA was trapped too quickly in a local optimum solution. However, this number was subject to the population size.
- Too low crossover probability preserves solutions to be interchanged and longer time is required to converge. This probability should be large enough to crossover almost all mated solutions.
- Too high mutation probability introduced too much diversity and took longer time to reach an optimal solution. Too low mutation probability tended to miss some near-optimal solutions. Using the annealing strategy the effects of too high or too low mutation were overcome.

6.   EVALUATION

First, the efficiency of an existing universal motor was calculated. The power losses of this motor were 313W and the output power 731W. In the outline (Fig. 2a), the levels of magnetic flux density through the rotor/stator lamination are shown. The darkest gray color indicates areas with the highest level of magnetic flux density, which results in high iron losses (2.3T).

After several runs of the GA, a set of promising solution candidates was collected. For each candidate a finite-element numerical simulation followed by the calculation of the fitness value was performed. Most of the solutions show a significant reduction of power losses in comparison with losses in the existing motor. The best solution results in a power losses reduction of 28%, and defines a motor with power losses of 226W. The comparison of the magnetic flux densities in the existing and the optimized (Fig. 2b) motors shows a clear reduction of areas with the highest levels of magnetic flux density in the optimized motor.

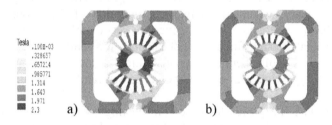

Figure 2. Lamination outline: a) initial, b) optimized

In the second phase of the evaluation the outer dimensions of the motor lamination were fixed to ensure unchanged (not increased) material costs and allow the implementation of the motor into the same housing as before. There is a difference (Tab. I) in power losses and efficiency of the optimal motor due to our additional constraints. However the overall losses were reduced by 16%. After a prototype of the motor was measured we could evaluate the real values of losses and its efficiency. These values are shown in Tab.I and are slightly different from the calculated ones, because of the mentioned non-exact iron losses calculation.

Table -1. Evaluation results

	analytic calculation		prototype measurement	
	existing	new	existing	new
input power	1044 W	1037 W	1100 W	1100 W
efficiency	70.0 %	74.5%	69.7%	72.6%
output power	731 W	773 W	767 W	799 W
power losses	313 W	264 W	333 W	301 W
efficiency difference		4.5%		2.9%
power losses difference		49 W		32 W

For both types of a motor the overall computation time was about 70 hours on Pentium III Computer Station. Here the population size was 30, number of generation was 100 while the crossover and mutation probabilities were 70% and 0.1% respectively.

7. CONCLUSION

In this paper we described an approach using the GA in a very early motor design phase when an optimal configuration of the geometrical parameters has to be found. The GA generates sets of solution candidates, which are evaluated using finite-element method. We demonstrated that by repeating the process of generating sets of solutions in the evolutionary way, an optimal configuration with reduced power losses can be found in a very short time.

Using the GA the power losses of an existing universal motor were reduced for at least 10% and up to 28%. Increasing the GA running time or setting its parameters more appropriately could still improve this result.

References

[1] T. Bäck, *Evolutionary Algorithms in Theory and Practice*, Oxford University Press, New York, 1996.

[2] *ANSYS User's Manual*, ANSYS version 5.6, 2000.

[3] C. L. Karr, I. Yakushin, K. Nicolosi, "Solving inverse initial-value, boundary-value problems via genetic algorithm," *Engineering Applications of Artificial Intelligence*, Vol. 13, No. 6, 2000, pp. 625-633.

[4] G. Papa, J. Šilc, "Automatic large-scale integrated circuit synthesis using allocation-based scheduling algorithm," *Microprocessors and Microsystems*, Vol. 26, No. 3, 2002, pp.139-147.

[5] B. Koroušić-Seljak, "Heuristic Methods for a Combinatorial Optimization Problem - Real-Time Task Scheduling Problem," *Smart Engineering System Design: Neural Networks, Fuzzy Logic, Evolutionary Programming, Data Mining, and Complex Systems*, ASME Press Series on Intelligent Engineering Systems through Artificial Neural Networks, Vol. 9, 1999, pp. 1041-1046.

A REVIEW OF THE DEVELOPMENT AND APPLICATION OF CLUSTER ORIENTED GENETIC ALGORITHMS

I. C. Parmee
Advanced Computation in Design and Decision-Making
Faculty of Computing, Engineering and Mathematical Sciences,
University of the West of England
Bristol, BS 16 1QY
UK
Ian.parmee@uwe.ac.uk
Iparmee@ad-comtech.co.uk

Abstract: Cluster Oriented Genetic Algorithms (COGAs) have the ability to identify high-performance (HP) regions of complex design spaces through the on-line filtering of solutions generated by a genetic algorithm [1]. COGAs support the designer by providing relevant information relating to the characteristics of HP regions. This can lead to the identification of best design direction during early stages of design or to reduce the complexity of design space through a reduction in variable range or the conversion of problem variables to fixed parameters both during conceptual and detailed design. The paper introduces the initial variable mutation cluster oriented genetic algorithm (vmCOGA) before briefly describing more recent improvements and their implications. Examples then follow of the utilisation of COGAs both for exploratory conceptual design and for variable space reduction during more rigorous stages of the design process.

Keywords: cluster oriented genetic algorithms, decision support, design space reduction.

1.   INTRODUCTION

The objective of COGA is the identification of high performance regions of a design space and the achievement of sufficient regional set-cover (in

T. Burczyński and A. Osyczka (eds),
IUTAM Symposium on Evolutionary Methods in Mechanics, 331–340.
© 2004 Kluwer Academic Publishers.

terms of number of solutions) to allow significant qualitative and quantitative information to be extracted. COGA comprises two primary components: the diverse search engine (DSE) which searches the design space and discovers regions of high performance and the adaptive filter (AF) which extracts and stores information relating to each identified region. Variable mutation COGA (vmCOGA) uses differing mutation rates to achieve an appropriate exploration / exploitation balance. A number of search stages, each consisting of any number of generations, are introduced. During initial stages, the mutation rate is high to ensure efficient sampling of the design space. In later stages the mutation rate is relaxed to promote convergence of search and formation of clusters of solutions in high performance regions. VmCOGA search is defined by two vectors: the generational vector (**g**) which defines the end point (in generations) of each search stage and the mutation vector (**m**) which defines the mutation probability used during each search stage.

$$
\mathbf{g}^{\mathrm{T}} = \begin{bmatrix} \textit{end of search stage 1 (gen}_1\textit{)} \\ \textit{end of search stage 2 (gen}_2\textit{)} \\ \textit{end of search stage 3 (gen}_3\textit{)} \\ \cdot \\ \cdot \\ \textit{end of search stage n (gen}_n\textit{)} \end{bmatrix} \quad \mathbf{m}^{\mathrm{T}} = \begin{bmatrix} \textit{p(mutation) search stage 1 (mp}_1\textit{)} \\ \textit{p(mutation) search stage 2 (mp}_2\textit{)} \\ \textit{p(mutation) search stage 3 (mp}_3\textit{)} \\ \cdot \\ \cdot \\ \textit{p(mutation) search stage n (mp}_n\textit{)} \end{bmatrix}
$$

The vectors, **g** = {5, 10, 15, 20, 25} and **m** = {0.08, 0.06, 0.04, 0.02, 0.01} therefore define a vmCOGA run of five search stages each consisting of five generations, the mutation probability during stage one to five is 0.08, 0.06, 0.04, 0.02 and 0.01 respectively.

The Adaptive Filter (AF) copies high fitness designs from the evolving population to the Final Clustering Set (FCS). The AF overcomes problems experienced by most multi-modal evolutionary algorithms relating to required apriori knowledge of the design space by modelling the fitness of the evolving population using the normal distribution. This probabilistic approach eliminates the need for apriori since the process maps the fitness distribution to the interval {0,1}. The filtering process itself may be further split into two processes, explicit and implicit filtering. Explicit filtering occurs after each search stage when each chromosome within the population is checked against a predefined filtering threshold (Rf). If the normalised fitness of the chromosome is greater than Rf, the solution is copied, with replacement, to the FCS (Figure 1). If the normalised fitness is less than Rf the solution does not enter the FCS. As with the diverse search engine, the

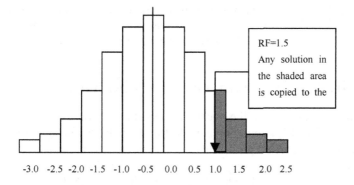

-3.0 -2.5 -2.0 -1.5 -1.0 -0.5 0.0 0.5 1.0 1.5 2.0 2.5

Normalised fitness distribution

Do (i=1 to i=POPSIZE)

{

$$f_i^n = \frac{f_i \square \alpha}{\square}$$

if $(f_i^n > Rf)$

{

　copy *i*th population

　member to FCS

}

}

f_i	=	fitness of *i*th population member
f_i^n	=	normalised fitness of *i*th population member
α	=	mean population fitness
$\square$	=	standard deviation
Rf	=	filtering threshold

Figure 1. The action of the adaptive filter

filtering thresholds applied during search are defined by a filtering vector, **Rf**,

$$\mathbf{Rf}^T = \begin{bmatrix} \text{filtering threshold used at the end of search stage 1} (Rf_1) \\ \text{filtering threshold used at the end of search stage 2} (Rf_2) \\ \text{filtering threshold used at the end of search stage 3} (Rf_3) \\ \cdot \\ \cdot \\ \text{filtering threshold used at the end of search stage n} (Rf_n) \end{bmatrix}$$

Hence, the filtering vector, **Rf** = {1.5, 1.5, 1.5, 1.5, 1.5}, defines five explicit filtering stages each using a filtering threshold of 1.5. Implicit filtering overcomes the loss of information between filtering stages by copying solutions to the FCS if their true fitness exceeds the fitness value associated with the previous filtering threshold [1].

The filtering process continually extracts high performance solutions during search without the need for apriori knowledge relating to the design model. A further advantage is the elimination of the need to maintain stable niches within the population since all solutions are extracted dynamically during search. During the early stages of search global solutions of relatively low fitness enter the FCS.   As search converges, solutions with higher fitness, located at or about optimal regions generally enter the FCS.

2.   IMPROVING SOLUTION GENERATION

Mutation alone is not the most efficient means for exploration and solution diversity. High mutation may destroy heritability during search and the setting of the mutation probability during search is by no means arbitrary [2,3]. Two novel exploration approaches have therefore been developed namely Halton Injection (HiCOGA) and Spatial Selection (SsCOGA).

HiCOGA increases design space sampling by injecting solutions generated from a low discrepancy sequence (LDS) directly into the crossover phase [4]. An LDS generates any number of points that uniformly fill an n-dimensional unit hypercube [5] ensuring more efficient search space sampling when compared with randomly generated points.   The Leaped Halton Sequence [6] is utilised. Solutions selected using objective function fitness are paired and reproduce with injected chromosomes. As the number of injected solutions increases, a greater proportion of diverse individuals take part in crossover, and the sampling of the search space increases. The number of injections made per generation may be used to attain a balance between exploration and exploitation (in the same way as mutation is employed in vmCOGA).   During the initial stages of search high numbers of injections ensures maximal sampling of the design space, as search progresses the number is relaxed to promote the convergence upon high performance regions of the search space.

Sampling within SsCOGA [4] utilises a dual selection scheme where increased potential for selection is placed upon more isolated individuals within a population.   Parents are selected using a local solution density metric and paired with further parents selected in terms of objective function fitness. For a given parent (P_1) its local density fitness (d_{nn}) is the Euclidean distance between itself and its nearest neighbour.   Using this metric, solutions from less densely populated regions will have an increased probability of being selected when compared with the solutions in more densely populated regions. As with HiCOGA, high numbers of spatial selections promote the maximal sampling of the design space, whilst reduced numbers result in lesser degrees of sampling and an increase in the rate of convergence upon regions of high performance within the design domain.

The ability of HiCOGA and SsCOGA to significantly increase sampling of the search space and to promote exploration of more sensitive regions has been demonstrated empirically and complete results can be found in [4] & [7].

The numbers of Halton injections and Spatial Selections are declared before and remain unaltered during search. Considerable emphasis is therefore placed upon the user who may not be familiar with either the search space under investigation or the COGA tool being used. To avoid this, the number of Halton injections and Spatial Selections needs to be controlled during search by some population convergence measure that is extracted from the evolving population. This approach eliminates the need for apriori tuning of both algorithms whilst also supporting the attainment and maintenance of optimal search capabilities during operation. A feedback metric specifically designed to increase the sampling of the search space uses a measure based upon the spatial diversity of the population. The metric is a development of the measure used to assign a spatial fitness to ssCOGA. Full details can be found in [7]. Results indicate higher levels of convergence within the FCS when compared with a test suite of algorithms including vmCOGA, HICOGA and SSCOGA.

3.   THE DYNAMIC ADAPTIVE FILTER

A drawback of both Spatial Selection and Halton Injection is an increase in the sensitivity of search. A Dynamic Adaptive Filter (DAF) has thus been developed to increase the robustness and accuracy of COGA by reducing the effects of two causes of poor filter performance, model drift and model mismatch [8].

Explicit filtering involves the normalisation of the population fitness distribution and its comparison against a predefined filtering threshold. During intermediate generations a solution is also copied to the FCS if its true fitness exceeds that associated with the previous filtering threshold (implicit filtering). Implicit filtering is therefore prone to a degree of inaccuracy (model drift) since the actual population mean and standard deviation change during the generations between explicit filtering..

Model mismatch occurs when the theoretical population distribution does not accurately model the true population. The problem of model mismatch will further augment the inaccuracies caused by model drift. Kurtosis and skewness are two scenarios where the actual population distribution may depart from that described by the normal distribution. If the degree of kurtosis becomes too great the normal probability distribution function (PDF) no longer adequately models the true fitness distribution and model mismatch occurs. As with kurtosis, if skewness becomes too great, model mismatch may result.

Model drift and mismatch are overcome by including explicit and implicit filtering into a single operation which occurs at the end of each generation after the first search stage. The DAF utilises a library of PDFs to model the population fitness distribution. The PDF that produces the closest match to the actual fitness distribution, according to a "closeness of fit" metric is then introduced. A probability mapping converts the value of Rf to an equivalent fitness threshold (f_{Rf}) for the chosen distribution. A population member is then copied to the FCS if the fitness exceeds the f_{Rf} threshold [8].

4.   THE COGA GRAPHICAL USER INTERFACE

COGA sits within a graphical user interface in which the user may set, monitor and control all COGA search parameters in addition to objective and variable bounds relating to the problem domain. The regions of high performance may then be explored by plotting any combination of two of the problem's design variables within a two dimensional hyperplane.

This is illustrated in Figure 2. The 2d hyperplane relating to gross wing plan area and wing aspect ratio variables of a BAE preliminary airframe design problem clearly illustrates the identification of a HP region relating to the maximisation of ferry range. Clicking upon any of the solutions within the HP region results in the ferry range value plus the vector of corresponding variable values appearing on the screen. Colour coding of the solutions gives a rapid indication of their relative fitness. The designer can immediately identify reduced lower limits of the bounds of these two variables in order to focus search in the high performance region. Calculating the standard deviation of the fitness of the solutions within the

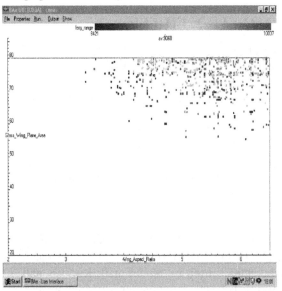

Figure 2. Selected design variables presented within a 2D hyperplane with solutions from the FCS projected upon it

FCS also gives an indication of solution robustness.

Figure 3 shows an alternative variable hyperplane with solutions from the same FCS projected upon it. In this case high performance solutions are

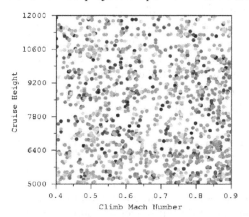

distributed across the entire variable ranges. The designer can, in this case take two courses of action:

□ Keep the variable bounds at their current settings;

□ Using the colour coding and the ability to pull up the fitness and variable vectors of any solution select a solution that satisfies requirements and fix the two variables defining the hyperplane to the values of the preferred solution.

Figure 3. Alternative hyperplane relating to different variables

The second option converts the two variable parameters to fixed parameters thereby reducing the dimensionality of the design space and its overall size.

Figure 4 shows the boundaries of three FCSs generated from individual COGA runs. The solutions in each relate to a differing objective of the same preliminary airframe design problem i.e. Ferry Range, Attained Turn Rate

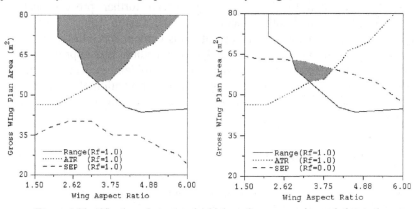

Figure 4. Identification of compromise high-performance regions relating to three differing objectives.

and Specific Excess Power.

Each bounded area represents a high-performance region relating to its specific objective. In the left-hand figure the same filtering factor (Rf) has been used for each objective. This projection of objective space onto variable space provides an indication of the probable degree of conflict between the objectives. The left-hand figure indicates that a common region for Ferry Range and Turn Rate has been identified but the Specific Excess Power objective cannot be satisfied with that filter setting. However, relaxing the Specific Excess Power filter threshold allows lower fitness solutions through and boundary moves as indicated in the right-hand figure resulting in the identification of a common region for all objectives. Relaxing the filter threshold could be considered to be a change in relative preference relating to the three objectives i.e. in order to satisfy all three we are willing to accept lower fitness solutions in one (or more).

5.   THE INTERACTIVE EVOLUTIONARY DESIGN STATION

COGA has been integrated with the Interactive Evolutionary Design Station (IEDS) [9,10]. The IEDS is a conceptual design / decision support tool comprising several components. At the IEDS core is a co-evolutionary component that simultaneously investigates differing aspects of the design problem. Agent-based software provides communication between evolutionary processes and between the user and the various IEDS modules. Designer preferences can be introduced and modified on-line by the user where such modification is based upon relevant information extracted from the co-evolutionary system. Information is continuously extracted by COGA and presented to the designer who can further process such information off-line with other members of the design team. COGA therefore contributes to the information-gathering component of the IEDS.

The aim of the IEDS is to provide a dynamic medium that improves the design team's knowledge base relating to the problem at hand. This promotes continuous interaction resulting in the re-formulation of the design problem as uncertainty and ill-definition decreases in proportion to knowledge gained.

6.   FURTHER WORK

It is necessary to introduce some independent check to ensure that the compromise regions shown in the projections of Figure 3 are actually mutually inclusive. On-line analysis of the solution vectors of the Final Clustering Sets provide increased confidence in the graphical representations plus further define the extent of each high-performance region [11].

Current work is investigating the utilisation of evolutionary classifier systems (CS) for similar high-performance region identification [12]. The intention is to integrate the COGA and CS system approach in order to extract further information relating to region bounds and variable / objective interaction in the form of succinct rules.

COGAs have been largely been perceived as conceptual design tools that can support decision-making in the preliminary stages of design. However, with the rapid increases in computational capability it now seems feasible to introduce the COGA concept to more detailed design processes. The integration of COGA with low-resolution finite element analysis could result in significant search space reduction in terms of both reduced variable ranges and the conversion of variable parameters to fixed parameters. It is intended to investigate this approach more fully.

7.   DISCUSSION

The paper reviews a number of improvements to the basic variable mutation cluster-oriented genetic algorithm. The replacement of vmCOGA with the injection of Halton individuals (HiCOGA) and the use of spatial selection (SsCOGA) improves the degree of search space sampling and convergence within the FCS. The control mechanism eliminates the need to set injection and sampling rates. The Dynamic Adaptive Filter (DAF) increases algorithm robustness. Five library PDFs cover the envelope of projected population fitness distributions. The DAF significantly increases algorithm robustness without any discernible reduction in the sampling of the search space.

Examples of the utilisation of the COGA concept relating to both single and multi-objective satisfaction are shown and the integration of the concept with an Interactive Evolutionary Design System has been briefly described.

A more detailed history of the development of COGA, its utilisation and its integration with the IEDS can be found in [13]. Detailed results relating to recent developments can be found in Chris Bonham's referenced papers and in his PHD thesis [7]. The reader is also directed to reference [14] for more information relating to the Interactive Evolutionary Design System.

References

[1] Parmee, I. C. (1996). The Maintenance of Search Diversity for Effective Design Space Decomposition using Cluster Oriented Genetic Algorithms (COGAs) and Multi-Agent Strategies (GAANT). Proc. Adaptive Computing in Engineering Design and Control, University of Plymouth, UK, PP 128-138.

[2] Davis, L. D. (1989). Adapting Operator Probabilities in Genetic Algorithms. Proceedings of the 3rd International Conference on Genetic Algorithms, pp 61-69.

[3] Bäck, T. (1992). The Interaction of Mutation Rate, Selection and Self-Adaptation within a Genetic Algorithm. In Manner et. al. (ed.) Parallel Problem Solving from Nature, 2, pp 85-94.

[4] Bonham, C. R. & Parmee, I. C. (1999) An Investigation of Exploration and Exploitation within Cluster Oriented Genetic Algorithms (COGAs). Proceedings of the Genetic and Evolutionary Computation Conference, July 13-17, Orlando USA, pp 1491-1497.

[5] Sobol, I. M. (1967). On the Distribution of Points in a Cube and the Approximate Evaluation of Integrals". *Computational Mathematics and Mathematical Physics*, 7, (4), pp 784-802.

[6] Kocis, L., Whiten, W. J. (1997). Computational Investigations of Low Discrepancy Sequences. ACM Transactions on Mathematical Software, 23, 2, pp266-294

[7] Bonham C. R., 2000, Evolutionary Decomposition of Complex Design Spaces. PhD Thesis, University of Plymouth.

[8] Bonham, C. R. & Parmee, I. C. (2000). Improving the Robustness of COGA: The Dynamic Adaptive Filter, Proceedings of the 4th International Conference of Evolutionary Design and Manufacture (ACDM'00), Plymouth, UK, pp 263-274.

[9] Parmee I. C., Cvetkovic D., Bonham C., Packham I., (2001), Introducing Prototype Interactive Evolutionary Systems for Ill-defined Design Environments. *Journal of Advances in Engineering Software*, Elsevier, **32** (6);pp 429-441.

[10]Parmee I. C., Cvetkovic D. Watson A. H., Bonham C., (2000), Multi-objective Satisfaction within an Interactive Evolutionary Design Environment. *Evolutionary Computation*, **8** (2), pp 197-222.

[11]Packham I. S., Parmee I.C., (2000), Data analysis and visualisation of cluster-oriented genetic algorithm output. *International Conference on Information Visualisation – IV2000*, 19-21 July 2000, published by IEEE Computer Society.

[12]Bull, L., Wyatt, D. & Parmee, I. (2002), *Towards the Use of XCS in Interactive Evolutionary Design*, Proceedings of the Genetic and Evolutionary Computation Conference 2002.

[13]Parmee, I. (2001), *Evolutionary and Adaptive Computing in Engineering Design*, Springer-Verlag London.

[14]Parmee I. C. Improving Problem Definition through Interactive Evolutionary Computation. *Journal of Artificial Intelligence in Engineering Design, Analysis and Manufacture,* 16 (3), Cambridge Press (2002 - in press),.

GENETIC ALGORITHM OPTIMIZATION OF HOLE SHAPES IN A PERFORATED ELASTIC PLATE OVER A RANGE OF LOADS

Shmuel Vigdergauz

Research and Development Division, The Israel Electric Corporation Ltd., P.O.Box 10, Haifa 31000, Israel and

Department of Mathematics, Technion - Israel Institute of Technology, Haifa 32000, Israel

smuel@iec.co.il

Abstract: The in-plane elastic behavior of a thin perforated plate is studied in the optimization context of finding the hole shapes which minimize the stored strain energy . In the literature, the loading conditions are usually assumed to be fixed with no changes allowed. The resulting optimal structure may thus appear to be ill-designed for even small load variations which inevitably occur in actual practice. Here, we consider the more realistic case when the plate is subject to any biaxial load from the interval between a hydrostatic state and pure shear. This is a minimax problem in which the worst admissible load and the best solution are identified concurrently. The numerical results are partially based on the previous author works's [13, 14]

Keywords: microstructure, plates, open hole, biaxial loading, load uncertainty, genetic algorithm

1. INTRODUCTION

Flat perforated construction elements are widely used in engineering. In many instances, they may be modeled as a thin infinite plate weakened by a regular array of traction-free holes. Though difficult to achieve in practice, the regularity assumption allows for averaging the stress micro fields over the cell. This procedure gives statistically correct assessments of the effective (i.e. macroscopic) moduli of the plate. They relate averages of an arbitrary stress macro field to the induced average strains in the convenient form of the local Hooke's law. At these

341

T. Burczyński and A. Osyczka (eds),
IUTAM Symposium on Evolutionary Methods in Mechanics, 341–350.
© 2004 Kluwer Academic Publishers.

settings, the maximum of a particular effective modulus provides the
minimum energy response and hence the most structure rigidity on the
corresponding loading.

By definition, the moduli are independent of the applied stresses. However, they do depend on the hole shape which thus may be specially
designed for making the perforated structure as stiff as possible. In the
literature, this shape optimization problem was considered mostly for a
fixed loading or equivalently for a separate effective modulus. Under this
restriction, the bulk-optimal [3, 10, 7] and the shear-optimal contours
[13] were explicitly found by various analytical and numerical methods.
The results obtained show that each modulus is being optimized with
some sacrifice in the others. This may limit the utility of the designed
structures in the more practical case of the load uncertainties.

To overcome this, [4] have modeled the changing loading conditions as
a set of different loads specially tailored for each concrete application.
The authors proposed a minimax formulation of the optimization problem when the energy is first maximized over the loads in order to find the
"worst" load at which the optimal structure is then identified. This approach has made it possible to identify the optimal two-phase composite
which is stable to variations in the uniaxial loading. Topologically, the
class of admissible composites includes the high rank matrix laminates
described by [2]. Whilst providing the global energy minimum at all
loads, these multi-scale structures are technologically unavailable.

Here we consider only more credible 1-scale perforated structures though
the global minimum is unreachable in this narrowed set at any but a hydrostatic loading [1, 13]. We also suppose that the load set is formed
by two orthogonal axial load with their ratio varying equiprobably from
one (hydrostatic loading) to minus one (pure shearing). These settings
permit a minimax formulation of the optimization problem when the
energy is first maximized over the loads in order to to find the "worst"
load at which the optimal structure is then identified numerically by the
genetic algorithm (GA).

Here, the standard GA is specially tailored for our purposes as was detailed in [12, 13]. First the Kolosov-Muskhelishvili periodic potentials
are combined in a fresh manner to perform a fast fitness evaluation. Second, an effective self-adjusting scheme is proposed to encode randomly
generated hole shapes with "automatic" retention of a given volume fraction $c < 1$. This technique helps to achieve a stable solution up to the
values $c \leq 0.85$.

The main conclusion obtained is that the pure shear is the "worst" admissible load at any hole volume fraction and hence the shear-optimal
contour is likewise optimal over the above described load set.

2. ANALYTICAL BACKGROUND FOR EVALUATING THE ELASTIC BEHAVIOR OF PERFORATED PLATES

Topologically, a perforated regular structure is formed by replicating a basic cell in two dimensions. Let the cell contain only one strictly interior centered hole with a smooth traction-free boundary L and the volume fraction $0 \leq c < 1$. For definiteness, let also the structure have two-fold (square) rotational symmetry which corresponds to the simplest anisotropic model with three different effective moduli: the bulk modulus K_e and two shear moduli μ_e, μ_e^*. The energy density W of any micro stress field $\{\sigma_{xx}, \sigma_{yy}, \sigma_{xy}\}$ is then written (see, for instance, [2]) as a positive definite quadratic (PDQ) form :

$$4W = \frac{Tr^2}{K_e} + \frac{Dev^2}{\mu_e} + \frac{\langle \sigma_{xy} \rangle^2}{\mu_e^*} \tag{1}$$

in the basic stress averages $\langle \bullet \rangle$ over the cell: $Tr \equiv \langle \sigma_{xx} \rangle + \langle \sigma_{yy} \rangle$, $Dev \equiv \langle \sigma_{xx} \rangle - \langle \sigma_{yy} \rangle$ and $\langle \sigma_{xy} \rangle$. They also serve as trial loadings to derive the effective moduli as the cell harmonic means: [11] :

$$\frac{1}{K_e} = \frac{1}{K} + \frac{4A_1}{E}; \ Tr = 1, \ Dev = \langle \sigma_{xy} \rangle = 0 \tag{2a}$$

$$\frac{1}{\mu_e} = \frac{1}{\mu} + \frac{4A_2}{E}; \ Dev = 1, \ Tr = \langle \sigma_{xy} \rangle = 0 \tag{2b}$$

$$\frac{1}{\mu_e^*} = \frac{1}{\mu} + \frac{4A_3}{E}; \ \langle \sigma_{xy} \rangle = 1, \ Tr = Dev = 0 \tag{2c}$$

Here K, μ and $E = 4K\mu/(K + \mu)$ are the bulk, shear and Young local moduli of the solid phase. Hexagonal structures with three-fold symmetry are macro isotropic [8] : $\mu_e = \mu_e^*$.
Given any geometry, the dimensionless coefficients $A_{1,2,3}$ are to be found by solving the direct doubly periodic elastostatic problem [11]. They describe the hole-induced changes in the local moduli K, μ and hence in the energy density W. With (1, 2) the density increment ΔW takes the form involving only one rather than two local elastic moduli:

$$4\Delta W = Tr^2 \left(\frac{1}{K_e} - \frac{1}{K} \right) + Dev^2 \left(\frac{1}{\mu_e} - \frac{1}{\mu} \right) + \langle \sigma_{xy} \rangle^2 \left(\frac{1}{\mu_e} - \frac{1}{\mu} \right)$$

$$= 4A_1 \frac{Tr^2}{E} + 4A_2 \frac{Dev^2}{E} + 4A_3 \frac{\langle \sigma_{xy} \rangle^2}{E} \tag{3}$$

At given structure symmetry type $A_{1,2,3}$ depend only on the hole shape and its volume fraction: $A_j = A_j(L, c), j = 1, 2, 3$ [11]. Table 1 gives the cellular (thin-walled) solid asymptotics ($c \to 1$) of the effective moduli and hence of their dominating coefficients $A_{1,2,3}$ found in [6] from the different considerations.

Of particular interest is a square cell under no twisting load : $\langle \sigma_{xy} \rangle = 0$

Table 1. The limit behavior of the coefficients $A_{1,2,3}$ from (2) with $c \to 1$

Cell type	A_1	A_2	A_3
Square	$\dfrac{c}{1-c}$	$\dfrac{c}{1-c}$	$\dfrac{4c}{(1-c)^3}$
Hexagon	$\dfrac{c}{1-c}$	$\dfrac{2c}{3(1-c)^3}$	$A_3 = A_2$

when the remaining items $A_{1,2}$ in (3) are of the same asymptotic order and hence may be compared indiscriminately in the whole interval $0 \leq c < 1$.

3. STATEMENT AND ANALYSIS OF THE OPTIMIZATION PROBLEM

Without loss of generality assume that one of two average loads, say, $\langle \sigma_{xx} \rangle$ is always non-zero, so that

$$Tr = (1+q)\langle \sigma_{xx} \rangle; \quad Dev = (1-q)\langle \sigma_{xx} \rangle$$
$$q = \langle \sigma_{yy} \rangle / \langle \sigma_{xx} \rangle; \quad -1 \leq q \leq 1$$

With this in view, we use Table 1 to normalize (3) as a PDQ form in q:

$$4w(L,c,q) \equiv (1-c) E \Delta W / \!\langle \sigma_{xx} \rangle^2 / c = \alpha_1(L,c)(1+q)^2 + \alpha_2(L,c)(1-q)^2 \quad (4)$$
$$\alpha_j(L,c) = (1-c)A_j(L,c)/c, \quad j = 1,2; \quad 0 \leq c < 1; \quad -1 \leq q \leq 1$$

In particular

$$w(L,c,1) = \alpha_1(L,c); \qquad w(L,c,-1) = \alpha_2(L,c) \qquad (5)$$

Now, suppose that the plate may be subject equiprobably and independently to any non-twisting load; that is the parameter q varies uniformly between minus and plus one. Then, (4) serves as the optimality criterion in the following minimax problem:

At given c, to find the square symmetric hole shape which minimizes the normalized energy density w:

$$\max_{-1 \leq q \leq 1} w(L,c,q) \xrightarrow[\{L\}]{} min \qquad (6)$$

over the range
$$-1 \leq q \leq 1 \tag{7}$$

As a PDQ function (4) of q, the criterion w has a maximum only at either of the interval endpoints $q = \pm 1$

$$\max_{-1 \leq q \leq 1} w(L, c, q) = \max[w(L, c, -1); w(L, c, 1)] \tag{8}$$
$$= \max[\alpha_2(L, c); \alpha_1(L, c)]$$

with both values in the brackets relating to the same hole shape L. Therefore, (6) is brought to the form

$$\min_{\{L\}} \max_{-1 \leq q \leq 1} w(L, c, q) = \min_{\{L\}} \max[\alpha_2(L, c); \alpha_1(L, c)] \tag{9}$$

As separate minimization criteria, the quantities (5) define the bulk- and shear-optimal holes, further referred to with the symbols K and μ, respectively:

$$\alpha_1(L, c) \xrightarrow[\{L\}]{} min \equiv \alpha_1^{(K)}(c) \tag{10a}$$

$$\alpha_2(L, c) \xrightarrow[\{L\}]{} min \equiv \alpha_2^{(\mu)}(c) \tag{10b}$$

The optimization problem (10a) was solved analytically [3, 10, 7] for any $c < 1$:

$$\alpha_1^{(K)}(c) \equiv 1 \tag{11}$$

in parallel with the parametric representation of the $K-$ (or equi-stress, [3]) shapes through doubly periodic functions. The more complicated criterion $\alpha_2^{(\mu)}(c)$ from (10b) is numerically found up to $c \leq 0.85$ [13] by a genetic algorithm (GA) approach. In doing so, the intuitive consideration was used by which the optimal contour may be only convex. This assumption, while not proved mathematically, keeps the GA from performing numerically unstable calculations early in the process. Other GA characteristics were chosen as given in Table 2. The limiting case of an isolated hole ($c \to 0$) has been pursued separately [9, 5, 12] with the numerical schemes involving the easier-to-compute Laurent series rather then doubly periodic expansions. It is worth noting that the $K-$holes are smooth, whereas the $\mu-$ contours have angular points in which the tangential stress changes its sign.

After finding the optimal shapes, their non-optimal complementary parameters $\alpha_1^{(\mu)}(c)$ and $\alpha_2^{(K)}(c)$ are also computed by solving the direct elastostatic problem [13] . All four moduli are shown in Fig. 1. In conformity with Table 1, the curves approach unit value from above at $c \to 1$.

Table 2. The GA operator types and their probability rates typically used in further optimizations

GA Parameter	Parameter value(s)
Gene	Integer number [0; 255]
Individual	Interface shape
Population size	800
Number of genes	91 for a square , 61 for a hexagon
Initial population	800 random individuals
Selection	Tournament
Elitism	Four best individuals
Crossover	1-point
Crossover rate	0.90
Creep mutation	By randomly changing a bit
Creep mutation rate	0.35
Jump mutation	By adding a random integer value typically [-4; 4]
Jump mutation rate	0.35
Stopping criterion	After 1200 iterations

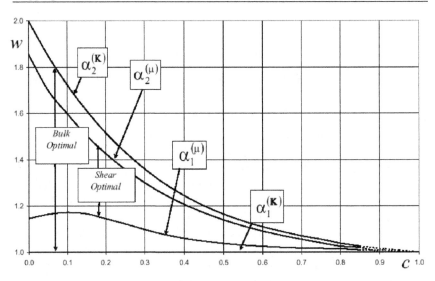

Figure 1. End-of-interval (extremal) values of the energy density w for the $K-$ and $\mu-$ square symmetric holes against the hole volume fraction $0 \le c \le 0.85$

4. THE NUMERICAL COMPARISONS BETWEEN THE EFFECTIVE MODULI OF THE BULK- AND SHEAR-OPTIMAL STRUCTURES

We are now in a position to compose the resolving inequalities for the considered optimization problem. Indeed, Fig. 1 suggests that for any

$c < 1$

$$w^{(\mu)}(c, -1) \equiv \alpha_2^{(\mu)}(c) > \alpha_1^{(\mu)}(c) \equiv w^{(\mu)}(c, 1) \tag{12}$$

Then, combining (8), (10b) and (12) yields the chain of relations which hold for any c and convex L

$$\max[w(L, c, -1); w(L, c, 1)] \geq w(L, c, -1) \geq \tag{13}$$

$$\min_{\{L\}} w(L, c, -1) = \alpha_2^{(\mu)}(c) = \max[\alpha_2^{(\mu)}(c); \alpha_1^{(\mu)}(c)]$$

Finally, we take the minimum in (13) and use (9) to prove the optimality of the μ–contours over the load range (7)

$$\min_{\{L\}} \max_{-1 \leq q \leq 1} w(L, c, q) = \alpha_2^{(\mu)}(c) \tag{14}$$

By (14) is meant that a pure shearing ($q = -1$) is the "worst" trial load in the interval (7) and hence the shear-optimal structures (10b) remain optimal in the more general sense of (6).

On the contrary, the K–structures (10a) are not optimal over the whole range (7) since their endpoint energies $w^{(K)}(c, -1)$, $w^{(K)}(c, 1)$ are in the relation

$$w^{(K)}(c, -1) \equiv \alpha_2^{(K)}(c) > \alpha_1^{(K)}(c) \equiv w^{(K)}(c, 1) \tag{15}$$

$$0 \leq c < 1$$

opposite to (12). The elastic responses of the both structures to different external loads are shown in Fig. 2. The depicted parabolic curves demonstrate that the bulk-optimal structures lose the over-the-range optimality only due to their elastic behavior under dominating shear stresses or, equivalently, in the interval $[-1; q_0(c))$ above the marked level $\alpha_1^{(K)} \equiv 1$. Here, the function $q_0(c)$ defines the point at which the energy difference $w^{(K)}(c, q) - w^{(K)}(c, 1)$ changes its sign from plus to minus:

$$w^{(K)}(c, 1) \equiv \alpha_1^{(K)} = w^{(K)}(c, q_0(c)) \tag{16}$$

$$-1 < q_0(c) < 1; \quad 0 \leq c < 1$$

Since $w^{(K)}(c, q)$ is a PDQ form (4) in q

$$4w^{(K)}(c, q) = \alpha_1^{(K)}(c)(1 + q)^2 + \alpha_2^{(K)}(c)(1 - q)^2 \tag{17}$$

(16) means that

$$\alpha_1^{(K)} \geq w^{(K)}(c, q); \quad q \geq q_0(c) \tag{18}$$

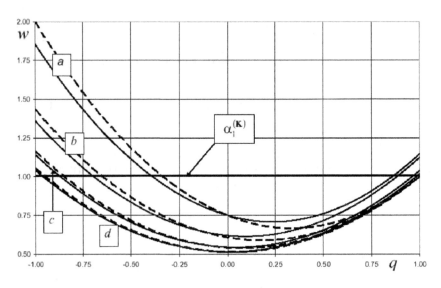

Figure 2.   The normalized energy increment $w(q, c)$ as a quadratic function of the biaxial ratio q for the hole volume fraction $c = 0.0(a),\ 0.25(b),\ 0.50(c),\ 0.75(d)$. The solid lines and the dashed lines correspond to the $\mu-$ and the $K-$structures, respectively

or, equivalently,

$$\alpha_1^{(K)} = \max_{q_0(c)\leq q\leq 1} w^{(K)}(c, q); \quad 0 \leq c < 1 \tag{19}$$

Then, again as before, we combine (19) with (8) and (10a) to derive the following inequalities

$$\max_{q_0(c)\leq q\leq 1} w(L, c, q) = \max[w(L, c, q_0(c)); \tag{20}$$

$$w(L, c, 1)] \geq w(L, c, 1) \geq \min_{\{L\}} w(L, c, 1) = \alpha_1^{(K)}(c)$$

$$= \max[\alpha_1^{(K)}(c); w^{(K)}(c, q_0(c))]$$

The $K-$shapes optimality over the narrowed load interval $[q_0(c); 1]$ is then proved by taking the minimum in both sides of (20):

$$\min_{\{L\}} \max_{q_0(c)\leq q\leq 1} w(L, c, q) = \alpha_1^{(K)}(c)$$

It only remains to derive the function $q_0(c)$ explicitly from (16). With (17) and (11) we have

$$q_0(c) = \left(\alpha_2^{(K)}(c) - 3\right) / \left(\alpha_2^{(K)}(c) + 1\right) \tag{21}$$

The relationship (21) is plotted in Fig. 3

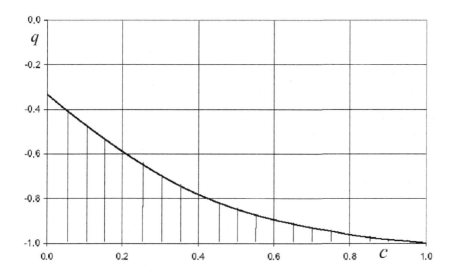

Figure 3. The left end $q_0(c)$ of the optimality interval $[q_0(c); 1]$ for the K–shapes. The hatching indicates the area where the separation inequality (18) does not hold

References

[1] Allaire, G.: Aubry, S., 1999: On optimal microstructures for a plane shape optimization problem *Struct. Opt.* (17)(2/3), 86–94.

[2] Avellaneda, M. 1987: Optimal bounds for elastic two-phase composites *SIAM J. Appl. Math.* (47), 1216–1228.

[3] Cherepanov, G.P.1974: Inverse problem of the plane theory of elasticity *J. Appl. Math. Mech* (38) (5), 913–931.

[4] Cherkaev, A.V.: Cherkaeva, E.A., 1994: Stable optimal design for uncertain loading conditions. *Homogenization : in memory of Serguei Kozlov.* Volume 50 of Series on advances in mathematics for applied sciences. Singapore: World Scientific.

[5] Cherkaev, A.V.: Grabovsky, Y.; Movchan, A.B.; Serkov, S.K. 1998: The cavity of optimal shape under shear stresses. *Int. J. Solids Struct.* (35) (33), 4391–4410.

[6] Gibson, L.J.; Ashby, M.E. 1999: *Cellular solids: structure and properties,* 2nd *edition.* Cambridge University Press

[7] Grabovsky, Y.; Kohn, R.V. 1995: Microstructures minimizing the energy of a two phase elastic composite in two space dimension. Part 2: The Vigdergauz microstructure. *J Mech. Phys. Solids* (43), 949–972

[8] Landau, L.D.; Lifshitz, E.M., 1986: *Theory of Elasticity.* Pergamon Press, Oxford.

[9] Vigdergauz, S.B.; Cherkaev, A.V. 1986: A hole in a plate optimal for its biaxial extension-compression. *J. Appl. Math. Mech.* (50) (3), 401–404

[10] Vigdergauz, S. 1986: Effective elastic parameters of a plate with a regular system of equal-strength holes. *Mech. Soilds* (21), (2), 162–166

[11] Vigdergauz, S. 1999: Complete elasticity solution to the stress problem in a planar grained structure. *Math. Mech. Solids* (4) (4), 407–441

[12] Vigdergauz, S. 2001a: Genetic algorithm perspective to identify energy optimizing inclusions. *Int. J. Solids. Struct.* (38) (38–39), 6851–6867.

[13] Vigdergauz, S. 2001b: The effective properties of a perforated elastic plate. Numerical optimization by genetic algorithm. *Int. J. Solids. Struct.* (38) (48–49), 8593–8616.

[14] Vigdergauz, S. 2002: Genetic algorithm optimization of the effective Young moduli in a perforated plate *Struct. Opt.* (24) (2), 106-117.

AN OBJECT ORIENTED LIBRARY FOR EVOLUTION PROGRAMS

with Applications for Partitioning of Finite Element Meshes

Jarosław Żola, Łukasz Laciński, Roman Wyrzykowski
Institue of Computer and Information Sciences
Technical University of Czestochowa, 42-200 Czestochowa, Poland
{zola,latin,roman}@k2.pcz.czest.pl

Abstract: In this paper, we present an object oriented library for evolution programs, developed at the Technical University of Czestochowa. The presented package contains number of C++ classes which allow to create various data structures and algorithms for evolutionary computations. This library supports optimized kernel and flexible user interface. Its main features are illustrated by the example of application to the problem of mesh partitioning.

Keywords: evolutionary programs, mesh partitioning, object oriented programming

1.   INTRODUCTION

Evolution programs (EPs) [3, 8], which group different search methods based on the mechanisms known from the world of biology, are one of the most interesting optimization procedures. The number of problems successfully solved using EPs is still increasing, as is the number of problem-solving toolkits based on EPs. Despite such interest in the application of these algorithms, the number of packages dedicated to creation of EPs is still rather small [8], perhaps because there are many requirements, which should be taken into account when such software is being developed. On the other hand, evolution of the object oriented languages, for example C++, provides developers with a number of functions allowing creation of more and more exact software models of the problems being analyzed. Generally, object oriented programming (OOP) [2] allows increased efficiency during software development, and

351

T. Burczyński and A. Osyczka (eds),
IUTAM Symposium on Evolutionary Methods in Mechanics, 351–360.
© *2004 Kluwer Academic Publishers.*

once software is created, OOP aids in reuse and extension by other developers and users.

In this paper, we present a C++ object oriented library for the evolution programs, which has been developed at the Institute of Computer and Information Sciences of the Technical University of Czestochowa. This library is an effect of studies on the application of genetic algorithms (GAs) in the field of the finite element analysis [13], but it can also be applied to optimization problems of different kind. In this paper, we show results of application of our library to the problem of mesh partitioning.

Our paper is organized as follows. In Section 2 we provide basic concepts of EPs, which determine the architecture of the library, presented in Section 3. The detailed description of our package is included in Section 4. In Section 5 we introduce the mesh partitioning problem, while in Section 6 we describe application of our library to this problem. Section 7 presents final conclusions.

2.   EVOLUTION PROGRAMS

The evolution programs are group of optimization procedures based on the mechanics of natural selection and inheritance. For the last few years, they are increasingly being applied as an effective solution for even very hard optimization problems. Despite EPs differ from each other, it is possible to distinguish only a few elements which are common for these methods.

Generally, we can say that evolution programs process a population, which is a set of coded potential solutions, using three genetic operators: selection, crossover and mutation [3, 8]. There are some exceptions of course, for example, the compact genetic algorithm [4] does not use population explicitly, and evolution strategies [8] utilize only mutation operator working with just one individual.

The important difference between various EPs can be found in the manner of genes and chromosomes representation. GAs use binary representation, while evolutionary algorithms (EAs) work with integers or floating point numbers. Some other techniques can deal with more complex structures. Another important issue arises from diversity of a particular genetic operator. Very often it is necessary to design a specialized version of a given operator, which depends on the constraints which occur in the considered problem. A good example of using this approach is partially matched crossover [8], which was designed for the traveling salesmen problem.

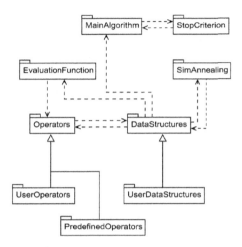

Figure 1. UML diagram of TEPLib library.

The relevant part of all EPs is the fitness function, which is used by the selection operator to evaluate a single solution, and always returns a positive value. Usually, the fitness function is built by conversion of the objective function. Additionally, in many cases it is scaled to improve efficiency of evolution program.

The last element of EPs, which is common to all optimization methods, is the termination condition. The evolution programs belong to the group of iterative procedures, so it is possible to apply criteria typically used by such methods, for example, fixed number of iterations. However, such an approach requires some knowledge of features of the objective function. Alternative methods are based on the analysis of the population and are more general.

3.   LIBRARY FOR EVOLUTION PROGRAMS

As it was discussed in the previous section, there are many functionality requirements which should be taken into account when software implementation of EPs is developed. Thanks to the object oriented approach [2], especially application of C++ with the STL library [10], it was possible to create a package covering most of the mentioned issues. Fig. 1 presents the UML diagram of our library, called *TEPLib*, which is abbreviation of *Templates Evolution Programs Library*.

The kernel of this library (see Fig. 2) consists of three packages: *DataStructures*, *Operators* and *EvaluationFunction*. Such a decomposition allows for an easy access to the main elements of evolution pro-

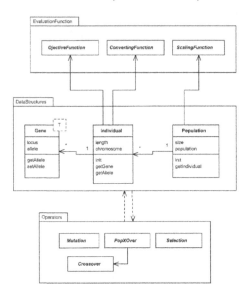

Figure 2.   UML diagram of TEPLib kernel.

grams, which is essential for the package functionality. It provides also two mutually independent parts: an algorithmic one, which consists of the genetic operators and the evaluation function, and the part containing data structure components.

These packages contain a number of template classes which describe basic EPs data structures, and define skeletons of main genetic operators. All these classes, for example *Population*, are based on the STL structures. As a result, the library uses efficient STL memory management and some of its generic algorithms. The application of classes from the packages *Operators* and *EvaluationFunction* to data stored in *DataStructures* is organized in a way which guarantees correctness of resulting user algorithms.

The kernel of the library is extended by another two packages, *MainAlgorithm* and *StopCriterion*, which allow complete evolution programs to be implemented.

The library contains also *SimAnnealing* package, which supports implementation of a simulated annealing algorithm. This package utilizes the *EvaluationFunction* module, and can easily replace *MainAlgorithm* (without reimplementation of the objective function). Additionally, the library includes a number of predefined genetic operators, which extend capabilities of our library.

4. LIBRARY DESIGN DETAILS

The most important part of the kernel is *DataStructures* package. There are three classes implemented in this module. The first class is *Gene*, which allows the user to describe a single gene, its allele, locus and gene domain (minimal and maximal value of gene). The type of gene is defined by a template parameter *T*. This class contains also the virtual method *flip* which provides the basic mutation procedure.

The next class is *Individual*, which describes a coded solution. This class is based on the STL *vector* class storing objects of type *Gene*. The use of the *vector* class makes initialization and modification of an individual easier. It also allows utilization of generic algorithms, like *fill* or *partition*. As a result, the user can create efficient crossover operators, using advantages of STL.

The *Population* class is a top-level structure describing the processed population. As in the previous case, it is an extension of the container of the *Individual* objects. This class offers a variety of methods, for example, statistic functions, useful in tracing behavior of the population.

Another three classes are included in the *EvaluationFunction* package. First of them is *ObjectiveFunction* class. This class is responsible for description of an objective function, and has to be implemented by user. This is possible by application of mechanisms of polymorphism and inheritance. The objective function is converted into the target function by application of *ConvertingFunction* class, which is a second class in the package. Finally, the last class, which is *ScalingFunction,* allows to implement a fitness scaling procedure.

The package *Operators* contains classes corresponding to genetic operators. An important feature of our design is that all these classes are pure virtual and represent only patterns, which have to be utilized when the final version of a given operator is created. However, the library includes a set of predefined (fully functional) examples of each operator. In addition to *Selection, Crossover* and *Mutation* classes, the *Operators* package offers another structure called *PopXOver*. This class is responsible for application of crossing to the whole population with a given probability. It allows different techniques of parents selection to be easily applied. All classes included in the *Operators* and *EvaluationFunction* packages are function objects, that is unary and binary functions.

5.   MESH PARTITIONING AND REFINEMENT

The finite element analysis (FEA) [12] techniques are often used to model physical phenomena. In FEA models, the physical system is represented as a system of differential equations, which is reduced to a linear system, created for a finite number of elements or nodes. The important issues concerning FEA (and other modeling methods) are time of computations and their accuracy. These two factors are strictly interrelated, and it is important to find a sensible proportion of them. The time of computations can be decreased by method parallelization [11, 12], and the increase of the result precision can be achieved by mesh refinement [1].

The parallelization of computations involves decomposition of the mesh into sub-domains (or sub-meshes) which are distributed over different processors. Following the classic formulation of partitioning [5], the partitioning of a mesh is performed before actual FEM computations. It should provide both equal or nearly equal computational load for all processors and minimized number of interfacing nodes. A low number of interfacing nodes means lower interprocessor communication, that is important for efficiency of parallel computations. Even in this formulation, the problem of mesh partitioning is NP-complete, so only heuristic methods of solving this problem are of practical value.

In our approach, the mesh generation and partitioning is made in two steps. First, for a given coarse background mesh the preliminary computations are run, and the mesh density is calculated. This parameter takes into account accuracy (or errors) of FEM computations, and can be calculated using some methods for checking quality of solution, for example, error estimation [6].

In the next step, a new mesh is generated using refinement of the coarse mesh, and the number of its nodes and elements is defined by the previously calculated density. The process of refinement can be performed concurrently over different processors. To achieve this goal, for each element of the coarse background mesh, the numbers of new nodes and elements, which will be generated, are estimated. The estimation is based on the mesh density. Then a partitioning is performed. The created sub-domains are distributed over target processors, which take care for refining the mesh.

The main advantage of our package is parallel mesh partitioning and generation based on the mesh density. As a result, memory bottlenecks of the sequential approach are avoided, since the resulting mesh is gen-

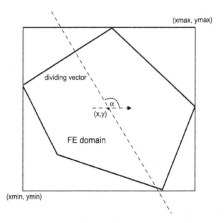

Figure 3. Bisection of finite element domain.

erated and stored on all processors. This permits the generation of very large meshes, which is normally impossible.

6. APPLICATION OF GA FOR MESH PARTITIONING

We have designed a genetic algorithm for the partitioning of 2D triangular finite element meshes, and implemented it using our library

The process of decomposition proceeds as follows. First, the estimation of the numbers of new elements and nodes is performed. This process is repeated for each element of the coarse mesh. For this purpose, the following expressions are used:

$$I = \left\lfloor \sqrt{\frac{\delta_i + \delta_j}{2}} \cdot l_{ij} \right\rfloor \tag{1}$$

$$f(e) = \left\lfloor \frac{1}{2} \cdot \left((\delta_i + \delta_j) \cdot l_{ij}^2 + (\delta_i + \delta_k) \cdot l_{ik}^2 + (\delta_j + \delta_k) \cdot l_{jk}^2 \right) \right\rfloor \tag{2}$$

where δ is density of the mesh in a given node, and l denotes distance between given nodes.

Formula (1) describes the expected number of new nodes which will be generated on the edge connecting nodes i and j of the background mesh. The second expression returns the expected number of new elements to be created in the coarse element e described by nodes i, j and k.

When the estimation is done, the background mesh is divided into two parts. This process is recursively repeated for each of resulting sub-meshes, until the number of returned sub-domains is equal to the number of target processors. The idea of a simple bisecting procedure is

tour. size	coding	part I	part II	part III	part IV	disbalance	interf. nodes
2	Gray	20705	20550	21019	20502	352	292
15	Gray	20514	20742	20775	20634	152	227
15	binary	20702	20583	20956	20383	300	271
10	Gray	20943	20392	20566	20587	321	274

Table 1. Results of partitioning with different parameters of GA.

presented in Fig. 3. The background mesh is partitioned by a dividing vector, which is described by three variables: x, y and α, where x and y are point coordinates, and α is a slope.

The parameters of the dividing vector are searched using GA. The objective function, which merges both requirements of mesh partitioning, is given by the expression presented below:

$$F = h(E_1, E_2) + |d(E_1) - d(E_2)| \qquad (3)$$

where $h(E_1, E_2)$ is the total number of expected interfacing nodes, and $d(E_1)$, $d(E_2)$ are the numbers of new elements in created sub-meshes.

This expression takes the minimal value when the number of elements in both sub-domains is equal and the number of interfacing nodes is lowest, thus it fulfills the two goals described in the previous section. We assume that both criteria are equally important. Formula (3) is a base for the fitness function which is a vital part of any GA. For purpose of our algorithm, we have implemented the following fitness function:

$$fit = 2 \cdot d(E) - F = 2 \cdot d(E) - h(E_1, E_2) + |d(E_1) - d(E_2)| \qquad (4)$$

where $d(E)$ is the total number of new elements.

The genetic algorithm used is based on the tournament selection [9] and uniform crossover operators [8]. It uses Gray coding for chromosomes representation [7].

Table 1 presents results of partitioning with different parameters of GA. For a coarse mesh with 1051 elements and 611 nodes, the table gives the estimated numbers of new elements in resulting four sub-meshes (see Fig. 4), followed by the value of load disbalance b, as well as the estimated number of interfacing nodes. Here the load disbalance is defined as

$$b = \max(|\frac{d(E)}{n} - d(E_i)|), i = 1 \ldots n$$

where n is a number of partitions. All results are averaged over 10 runs (with very small variance).

The results in the first three rows were obtained using GA with the size of population equal 150. The crossover probability was 0.9 and

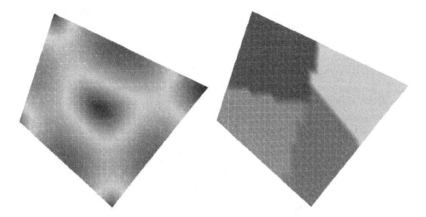

Figure 4.   Mesh density and example of partitioning.

mutation 0.01; each parameter of dividing vector was coded with 8 bits. The last row describes results of partitioning for the population size of 100 individuals. The crossover probability was 0.75 and mutation 0.01; each parameter of dividing vector was coded with 5 bits. In this case, the time of computation was approximately 30% shorter than in the previous cases. This effect was achieved with a loss of quality of partitioning.

As it is shown, the best results are obtained by the version of GA with the highest size of the tournament, and with the Gray coded population.

7.   CONCLUSIONS

In this paper, we have presented the package for evolution programs. Thanks to use of an object oriented approach, it allows evolution methods to be quickly implemented. Examples of its application for the problem of partitioning have been shown. Our library provides a flexible and extensible framework suited for use as a tool by other developers and users.

References

[1] M. Burghardt, L. Laemmer, U. Meissner, Parallel adaptive mesh generation, *Advances in Computational Mechanics with Parallel and Distributed Processing*, Saxe-Coburg Publ., Edinburgh, 1997, pp. 45-51.

[2] P. Coad, E. Yourdon, Object-Oriented Design, Prentice Hall, Inc., 1991.

[3] D.E. Goldberg, Genetic Algorithms in Search, Optimization and Machine Learning, Addison-Wesley, 1989.

[4] G.R. Harik, F.G. Lobo and D.E. Goldberg, The compact genetic algorithm, Tech. Report 97006, Univ. of Illinois at Urbana Champaign, 1997.

[5] B. Hendrickson, T.G. Kolda, Graph partitioning models for parallel computing, *Parallel Computing*, vol. 26, 2000, pp. 1519-1534.

[6] L. Lacinski, R. Wyrzykowski, J. Kaniewski, Parallel Meshing Algorithm for Finite Element Modeling, *Proc. Int. Workshop on Parallel Numerics - ParNum 2000*, Bratislava, Slovakia, 2000, pp. 117-124.

[7] K.E. Mathias and D. Whitley, Transforming the search space with Gray coding, in: *IEEE Conf. on Evolutionary Computation*, vol. 1, 1994, pp. 513-518.

[8] Z. Michalewicz, Genetic Algorithms + Data Structures = Evolution Programs, Springer-Verlag, Berlin, 1996.

[9] B.L. Miller and D.E. Goldberg, Genetic algorithms, tournament selection, and the effects of noise, Tech. Report 95006, Univ. of Illinois at Urbana Champaign, 1995.

[10] D.R. Musser, G.J. Derge and A. Saini, STL Tutorial and Reference Guide: C++ Programming with the Standard Template Library, Addison-Wesley, 2001.

[11] B.H. Topping and A.I. Khan, Parallel Finite Element Computations, Saxe-Coburg Publ., Edinburgh, 1996.

[12] R. Wyrzykowski, N. Sczygiol, T. Olas and J. Kanevski, Parallel finite element modeling of solidification process, *Lect. Notes in Comp. Sci.*, vol. 1557, 1999, pp. 183-195.

[13] J. Zola, R. Wyrzykowski and L. Lacinski, Using genetic algorithms for mesh smoothing and partitioning, *Proc. 5th Conf. on Neural Networks and Soft Computing*, Zakopane, Poland, 2000, pp. 663-668, 2000.

Mechanics

SOLID MECHANICS AND ITS APPLICATIONS
Series Editor: G.M.L. Gladwell

Aims and Scope of the Series

The fundamental questions arising in mechanics are: *Why?*, *How?*, and *How much?* The aim of this series is to provide lucid accounts written by authoritative researchers giving vision and insight in answering these questions on the subject of mechanics as it relates to solids. The scope of the series covers the entire spectrum of solid mechanics. Thus it includes the foundation of mechanics; variational formulations; computational mechanics; statics, kinematics and dynamics of rigid and elastic bodies; vibrations of solids and structures; dynamical systems and chaos; the theories of elasticity, plasticity and viscoelasticity; composite materials; rods, beams, shells and membranes; structural control and stability; soils, rocks and geomechanics; fracture; tribology; experimental mechanics; biomechanics and machine design.

1. R.T. Haftka, Z. Gürdal and M.P. Kamat: *Elements of Structural Optimization.* 2nd rev.ed., 1990
    ISBN 0-7923-0608-2
2. J.J. Kalker: *Three-Dimensional Elastic Bodies in Rolling Contact.* 1990 ISBN 0-7923-0712-7
3. P. Karasudhi: *Foundations of Solid Mechanics.* 1991   ISBN 0-7923-0772-0
4. *Not published*
5. *Not published.*
6. J.F. Doyle: *Static and Dynamic Analysis of Structures.* With an Emphasis on Mechanics and Computer Matrix Methods. 1991   ISBN 0-7923-1124-8; Pb 0-7923-1208-2
7. O.O. Ochoa and J.N. Reddy: *Finite Element Analysis of Composite Laminates.*
    ISBN 0-7923-1125-6
8. M.H. Aliabadi and D.P. Rooke: *Numerical Fracture Mechanics.*   ISBN 0-7923-1175-2
9. J. Angeles and C.S. López-Cajún: *Optimization of Cam Mechanisms.* 1991
    ISBN 0-7923-1355-0
10. D.E. Grierson, A. Franchi and P. Riva (eds.): *Progress in Structural Engineering.* 1991
    ISBN 0-7923-1396-8
11. R.T. Haftka and Z. Gürdal: *Elements of Structural Optimization.* 3rd rev. and exp. ed. 1992
    ISBN 0-7923-1504-9; Pb 0-7923-1505-7
12. J.R. Barber: *Elasticity.* 1992   ISBN 0-7923-1609-6; Pb 0-7923-1610-X
13. H.S. Tzou and G.L. Anderson (eds.): *Intelligent Structural Systems.* 1992
    ISBN 0-7923-1920-6
14. E.E. Gdoutos: *Fracture Mechanics.* An Introduction. 1993   ISBN 0-7923-1932-X
15. J.P. Ward: *Solid Mechanics.* An Introduction. 1992   ISBN 0-7923-1949-4
16. M. Farshad: *Design and Analysis of Shell Structures.* 1992   ISBN 0-7923-1950-8
17. H.S. Tzou and T. Fukuda (eds.): *Precision Sensors, Actuators and Systems.* 1992
    ISBN 0-7923-2015-8
18. J.R. Vinson: *The Behavior of Shells Composed of Isotropic and Composite Materials.* 1993
    ISBN 0-7923-2113-8
19. H.S. Tzou: *Piezoelectric Shells.* Distributed Sensing and Control of Continua. 1993
    ISBN 0-7923-2186-3
20. W. Schiehlen (ed.): *Advanced Multibody System Dynamics.* Simulation and Software Tools.
    1993   ISBN 0-7923-2192-8
21. C.-W. Lee: *Vibration Analysis of Rotors.* 1993   ISBN 0-7923-2300-9
22. D.R. Smith: *An Introduction to Continuum Mechanics.* 1993   ISBN 0-7923-2454-4
23. G.M.L. Gladwell: *Inverse Problems in Scattering.* An Introduction. 1993 ISBN 0-7923-2478-1

Mechanics

SOLID MECHANICS AND ITS APPLICATIONS
Series Editor: G.M.L. Gladwell

Mechanics

SOLID MECHANICS AND ITS APPLICATIONS
Series Editor: G.M.L. Gladwell

Mechanics

SOLID MECHANICS AND ITS APPLICATIONS
Series Editor: G.M.L. Gladwell

Mechanics

SOLID MECHANICS AND ITS APPLICATIONS
Series Editor: G.M.L. Gladwell

Mechanics

SOLID MECHANICS AND ITS APPLICATIONS

Series Editor: G.M.L. Gladwell

Kluwer Academic Publishers – Dordrecht / Boston / London